When the Fish Are Gone

When the Fish Are Gone

Ecological Disaster and Fishers in Northwest Newfoundland

Craig Palmer and Peter Sinclair

Fernwood Publishing • Halifax

Editing: Douglas Beall
Cover and inside photos: Craig Palmer
Design and production: Beverley Rach
Printed and bound in Canada by: Hignell Printing Limited

A publication of:
Fernwood Publishing
Box 9409, Station A
Halifax, Nova Scotia
B3K 5S3

Fernwood Publishing Company Limited gratefully acknowledges the financial support of the Ministry of Canadian Heritage and the Nova Scotia Department of Education and Culture.

Canadian Cataloguing in Publication Data

Main entry under title:

When the Fish Are Gone

Includes bibliographical references.
ISBN 1-895686-77-6

1. Fisheries -- Newfoundland -- History. 2. Fisheries -- Economic aspects -- Newfoundland. 3. Fisheries -- Social aspects -- Newfoundland. 4. Fishers -- Newfoundland -- Social conditions.
I. Sinclair, Peter R., 1947- II. Title.

SH224.N7P36 1997 338.3'727'0971809718 97-950033-8

Table of Contents

List of Illustrations

Maps

Tables

Figures

Preface

After more than a century of being inhabited by small-boat fishing communities with few social differences, northwest Newfoundland experienced rapid technological and social changes between 1965 and 1980. By 1982 a division was evident between the prosperous owners of a dragger fleet that used a type of mobile gear known as an "otter trawl" (a long, tapered net dragged on the ocean floor behind a vessel) and the vast majority of fishers, who were restricted by government regulations to using less productive stationary harvesting technologies. This technological division was reflected in a social division that pervaded life in the area, with the dragger fishers forming a local elite and the other residents resenting both this elite and the government for preventing them from becoming part of it.

In his book *From Traps to Draggers,* Peter Sinclair chronicled these events. That book focused on a theoretical debate concerning "the fate of domestic commodity producers within societies dominated by mature capitalist modes of production" (1985:17). Domestic commodity production (DCP) is a form of production in which the household provides most of the labour and other resources to produce goods that are marketed in order to sustain the household. Sinclair's book argued that none of the theories that predict either the rapid demise of DCP, its functional integration into the capitalist system or the social differentiation of domestic commodity producers into capitalists and proletarians had been formulated in enough detail to account for the structural factors operating in specific locations such as northwest Newfoundland. Hence, Sinclair concluded that

> [t]he implication of this case study for the general theory of differentiation is that, in areas with a narrow economic base, the intervention of the state can hold back, and even reverse to some extent, the economic pressure towards proletarianization of domestic commodity producers. At the same time, state policy in a capitalist society is unlikely to prevent the most efficient of these producers from expanding into more technologically sophisticated petty capitalist enterprises. Thus, only one dimension of the differentiation process was held back in the initial years (1985:144).

By the late 1980s the situation in northwest Newfoundland had been transformed by the growing realization that the cod stocks upon which most fishers depended were in serious danger of extinction. This growing ecological concern culminated in the closure of the cod fishery at the end of 1993, following the 1992 moratorium on Atlantic cod off northeast Newfoundland.

Although the events of the past thirteen years provide an opportunity to continue the analysis of the fate of domestic commodity producers in capitalist societies, the profound changes that have transpired since 1982 require a broader perspective. The apparent collapse of the cod stocks and the subsequent closure of the fishery require the integration of theories of DCP with theories concerning resource management. Hence, in this book we shall examine the interactions between the technological changes, governmental regulations, social divisions and ecological factors that have led to a situation where the continued habitation of northwest Newfoundland is itself in question. Although this investigation is focused on the coast of northwest Newfoundland, much of the analysis has obvious implications for the entire province of Newfoundland and all of Atlantic Canada. To a lesser degree, it is also relevant to the impact of environmental resource crises and the fate of DCP everywhere.

The material presented in this book draws upon many sources. In addition to studies on the ecology and economy of the fishery produced by federal and provincial governments and other agencies, we build upon the work of earlier research on social relations in the area. A detailed description of life in a small northwest Newfoundland outport during the pre-dragger era was provided by Firestone (1967). More recent sociological or ethnographic descriptions have also been produced by House et al. (1989), Felt and Sinclair (1995) and, for the eastern side of the Great Northern Peninsula (GNP), Omohundro (1994).

Our current work draws most heavily, however, on the research performed by the two authors. The description of the rise of the dragger fleet provided in Chapter 2 comes primarily from Sinclair's research in Port au Choix during 1981 and 1982. Some of the material on social and economic life on the GNP during the late 1980s comes from a major survey conducted by Sinclair and Felt in 1988. The description of social divisions and fishing strategies in reaction to declining cod stocks during the early 1990s is based primarily on Palmer's fieldwork in the area from 1990 to 1992. Reactions to the closure of the fishery and views of the future were gathered by Palmer and Sinclair during additional fieldwork in 1994. Please see the Appendix for details about our data collection methodology.

1
Introduction

What is the fate of "domestic commodity production" in capitalist economies? This is the fundamental question asked in this book. Our goal is to describe the fate of DCP in northwest Newfoundland in such a way that the general question is illuminated and the experiences of people in this particular location are better understood. What helps DCP survive and what pushes it to extinction? Northwest Newfoundland is an appropriate location to identify the factors that influence the fate of DCP because DCP was the dominant, if not exclusive, form of production in this area from 1905 to 1965. Previous studies (Sinclair 1985; Reinhardt and Barlett 1989) have indicated that differences in political, socio-cultural and ecological factors can produce fundamental differences in the fate of DCP, so here we will present a detailed analysis of how these factors have influenced northwest Newfoundland over the period from 1965 to 1995.

We will show how an ecological crisis has produced a social crisis: the small-boat fishery is in deep trouble, so are some of the petty capitalist vessels and so are fish plants and dependent communities. This analysis should be viewed in the context of the North Atlantic fisheries and, indeed, of all primary resource-producing rural areas in mature capitalist societies. Rural people are locked into a global economy they cannot expect to control and that they hope leaves them a niche in which to survive, if not prosper. Yet rural people are also incredibly inventive in creating adaptive strategies that allow space for local action within the structured environment of international commodity production (e.g., Long 1984; Felt and Sinclair 1995).

Small-scale agriculture producers are losing market share and social power to more capitalistic farms and agribusiness corporations (Friedland et al. 1991; Bonnano et al. 1994), while farm practices threaten the long-term productive potential of the land (e.g., Lind [1995] on rural Saskatchewan). Technological changes increase productivity but also raise investment requirements. Thus farmers with fewer resources are forced out of agriculture because they cannot sell their produce at prices high enough to allow them a decent living. The exodus from the land that began about 1940 in much of Canada led to a reduction of 60 percent in the number of farms between 1941 and 1986 (Hay 1992:25). Of those that remain, the bulk of production is concentrated in relatively few enterprises. For example, the top one percent of Canadian farms accounted for 19 percent of sales by 1981 (Winson 1992:91). It is also evident that the number of small, part-time or hobby farms is increasing. Thus what seems to be happening in many spheres of agriculture is that the "middle"—the family farm based on DCP—is seriously threatened (Buttel and LaRamee 1991). Often there is a sense of crisis among those still on the land and in small towns, a sense captured by this farmer activist from Saskatchewan: "I just look around town and see how tough times are, with all the houses vacant, people leaving friends that you knew and grew up with and all of a sudden, they are gone trying to make a go of it elsewhere" (Lind 1995:23). The impact of changes in the structure of agriculture is visible in many societies. As Friedland (1991:19) observes,

> there has been a growing recognition that the so-called agricultural "crisis" is structural and endemic to every national economy of advanced capitalism and exhibits continuity in a series of overproduction "crises" over more than a century.

In forest-based industries, major changes have also been under way. Timber production is dominated by large paper and sawmill companies, whether or not they own the forest land, at least in North America (Marchak 1983, 1991; Bailey et al. 1993, 1996). Here too, clearcutting practices and replantation in single species of trees create problems of erosion and destruction of natural habitat. Technological changes have transformed woods work as machines displace human labour. Moreover, the introduction of feller bunchers, skidders and large-scale road transportation to the mills has cut into the niche that small operators had previously filled (Reimer 1992; Marchak 1983; Bailey et al. 1996). Although wages and working conditions have improved for those still employed, especially in the pulp mills, change has come at a cost to DCP woods operators.

Environmental problems of varying degrees of severity and the transformation of the social organization of primary production through its integration into a corporate capitalist system are widespread issues. Fisheries have certainly been no exception, as many fish species and human populations that depend on them for food or employment have been in trouble around the globe in the 1980s and 1990s (McGoodwin 1990). North Atlantic herring, cod and capelin are certainly well-known examples of overfished species whose reductions have had serious social impacts (on the recent Norwegian case, see Jentoft [1993]).

Coming closer to our study area, on Canada's west coast, the salmon fisheries have been in crisis as the various interest groups, including states, struggle for control and Canadian producers try to work out their conflict with Americans from both north and south. Most of Atlantic Canada's groundfish producers have been hit hard by the closures of many stocks. The government hopes for a reduction of about 50 percent in the industry's labour force and is promoting change through the job retraining at the core of the Atlantic Groundfish Strategy (TAGS), a government program that provides support for displaced fishers and plant workers. Many boats have been tied up and plant doors closed, although shellfish, capelin and herring do provide some employment and even large incomes for those with appropriate licences and gear.

Many people believe they will be forced to move in the near future. Our research on the Bonavista Peninsula on the northeast coast revealed that more than 30 percent of those surveyed expected to be living elsewhere within five years. Among those under 30, the figure was 66 percent. This study was conducted late in 1994 when TAGS still had three years to run and the situation probably looked less desperate than it does today (Sinclair et al., forthcoming). Out-migration was being considered by over 30 percent on the Great Northern Peninsula in 1988 (Felt and Sinclair 1995). Today we have numerous anecdotal accounts that suggest migration is occurring on a large scale from many small fishing villages all over the province.

The Gulf coast fisheries off the Great Northern Peninsula provide a good example of the social and biological consequences of the failure to create conditions that would allow for fishing on a sustainable basis. This is also a matter on which we have collected extensive earlier research (Felt and Sinclair 1995; Firestone 1967; Omohundro 1994; Palmer 1992; Sinclair 1985), and thus it is especially attractive to bring that work up to date. What produced the problems in this area? How have local people attempted to carry on their lives? What are the implications for the social organization of the fishery in the future? Of course, no two situations and histories are precisely alike, but many areas have undergone similar experiences and, consequently, this case study should be of wide interest.

The book is organized according to the following framework. In this chapter, DCP is defined and the various theories of its fate in capitalist societies are summarized. Special attention is given to those few works that have attempted to apply theories of DCP to fishing communities. An overview of northwest Newfoundland and its history of DCP until 1965 is also presented. Chapter 2 describes the origin and growth of the small-scale, capitalist dragger fleet between 1965 and 1982. This includes an examination of the role of the state in differentiating area residents into a majority that remained locked into DCP and a minority of petty capitalists. Chapter 3 reviews the years 1982–87, when the state struggled to manage a dragger fleet that was flourishing but also beginning to threaten the fixed-gear fishery and, eventually, the health of the Gulf cod stock. Chapter 4 presents a detailed examination of the effect of these developments on the social relations of residents and discusses the role of social divisions in influencing state policy and determining the fate of DCP. Chapter 5 examines the collapse of the Gulf cod stock between 1988 and 1992. It focuses on the ecological factors underlying this collapse and on how the social construction of these factors influenced state policy. Chapter 6 analyzes the closure of the cod fishery in 1993. It investigates how this closure was influenced by the earlier closure of the northern cod stock in 1992 and how residents responded to the moratorium. Finally, Chapter 7 examines the likely impact of the previously described events on the future of DCP in northwest Newfoundland.

Our conclusion is that, despite several factors promoting its demise—including persistently low incomes, rising expectations of material affluence, lack of effective political pressure and the indefinite closure of the cod fishery—DCP has persisted in northwest Newfoundland. We will argue, however, that this persistence, far from being intended and desired by political planners, largely results from two unanticipated sets of events. The first, primarily socio-political, involves several failed attempts by the state to influence people to leave the DCP sector of the fishery. These attempts appear to have failed through poor planning and lack of enforcement. The second, primarily ecological, is essentially the ironic fact that the collapse of the Gulf cod stock may have interrupted policies leading to the reduction of DCP. Today, the future of DCP production is much more precarious than the number of active DCP fishers might indicate. To develop this analysis more fully, we should first explicate the concept of domestic commodity production.

Domestic Commodity Production

Other common terms with the same general meaning as DCP are *independent, simple and petty commodity production. Independent* implies that the producer is in control of the production process rather than subject to the control of an owner. It may imply a greater degree of autonomy than is reasonable in the common situation where the commodity producers cannot control prices and returns. *Simple* refers to the idea that this form of production reproduces itself but is not a source of capital accumulation or expanded reproduction. *Petty* implies a small scale but not insignificance. None of these terms can be considered wrong, but we feel that *domestic* best captures the key point to emphasize: the social organization of production.

The term *domestic* refers to the importance of kin, usually living in the same or neighbouring households, in supplying and organizing labour. The term implies that the household has no greater objective than the simple reproduction of the unit, as opposed to the expanded reproduction of petty capitalists. In his earlier work, Sinclair (1985) also distinguished between petty or small-scale capitalist production and DCP in a way similar to Llambi's (1988) analysis. Whereas DCP involves the labour of the owner and often other household members, petty capitalism combines the owner's labour with hired labour. These distinctions are useful because they point to different orientations and social interests among primary producers and help to illuminate processes of social change.

The term *commodity* refers to the fact that the household, although it may also be engaged in subsistence production, depends on the production of commodities and articulation with commodity markets to realize the value of what is produced and to acquire personal consumption goods and the means of production. In the case of northwest Newfoundland, this articulation is with a larger capitalist mode of production characterized by the private accumulation of capital appropriated from wage-labourers.

Occasionally, DCP is treated as synonymous with a peasant form of production, but it is better to avoid limiting the term to agriculture where it is often stretched to cover people who engage primarily in subsistence production. In general, the term *peasant* encompasses too wide a range of organizations and practices to be theoretically coherent, unless one of the narrow definitions (such as Friedmann's [1980] independent household production) can be effectively promoted.

The specific structure of the domestic commodity producing household in northwest Newfoundland is the patrilocal extended family (see Firestone 1967). This type of kin-based domestic structure is formed when sons continue to work and live with or near their father, even after reaching adulthood. In northwest Newfoundland, brothers were even expected to continue to work together and pool capital, labour and profits after the death of their father. This structure may owe its existence to the area's reliance on the cod trap which is uniquely suited to meeting the relatively high capital and labour demands necessary to perform this style of fishing (Firestone 1967). The patrilocal extended family also provided the large amount of labour involved in processing fish during this period.

Sinclair (1985) argued that analysis of fishing can draw from the much larger body of work on agriculture wherein there have been voluminous writings on the likely fate of DCP in capitalist societies. At the risk of oversimplification, authors can be allocated to one of three categories. The largest group identifies an historical movement of internal differentiation and proletarianization, allowing DCP to be perceived as a limited, transitory phenomenon. A smaller group presents broadly related views to the effect that DCP can persist either because it is insulated from capitalism or, more commonly, because its continuation is functional for capitalism. Finally, we encounter authors who are ambivalent regarding the final impact of structural forces, which are perceived to be operating in opposite directions.

Interpretations of Canadian fisheries seldom deal directly with the issues that concern us here. However, there are some relevant works. Clement (1986:61–82) presents an overview of the social organization of the fishery that is consistent with our perspective. His concern is to relate forms of production to different patterns of collective action, whether association, co-operative or union, but he does not address the question of historical trajectories. Neither the origin nor the future of these forms of production (including DCP) is analyzed.

Apostle and Barrett (1992) attempt to explain the decentralized structure of the Nova Scotian fishing industry, both harvesting and processing, which they see to be effectively resisting corporate capitalist attempts to take it over. Their analysis is located in the wider context of studies of the expansion of small capitalist enterprise in the twentieth century. This literature is relevant insofar as DCP can be considered to occupy a place similar to small-scale capitalist enterprises in economies dominated by large capital. Whether DCP enterprises will become petty capitalist is itself an important issue, but Apostle and Barrett miss this question by choosing to analyze small capital, which includes both; that is, their conceptual scheme does not permit the question to be asked. Apostle and Barrett (1992:22–30) favour a "diffusionist" perspective in which no inevitable concentration of capital is projected; small capital takes various organizational forms and has the potential to expand. Small firms tend to be more innovative and new products provide a market niche for them. Family sources of capital, paternalist relations in the firm and "community ties" are said to help small businesses succeed. This approach does not see small capital driven by or subject to a necessary dynamic of accumulation. Their review of the specific factors that make fishing unattractive to capital (including the inability to control access to the resource and the high risks entailed) leads them to the rather extreme conclusion that "any degree of capitalist centralization and concentration in the fishing industry is surprising" (Apostle and Barrett 1992:37). Barrett's chapter on post-war development continues the analysis. Of particular significance to this study is his specification of the factors that encourage small-scale fishing: market diversity, supportive state policy and technological developments that raise productivity.

Apostle and Barrett (1992) write as if small capital is prospering in the Nova Scotian fishing industry but they include diverse groups in this category, and these fragments of small capital may undergo different historical experiences. For example, both small, operator-owned draggers and Cape Island-style, coastal lobster and groundfish boats would qualify as small capital. In most instances, the first group is petty capitalist and the latter DCP. Based on research in Digby Neck and the islands in 1984, Anthony Davis (1991)[1] demonstrates that the rise of the small dragger fleet was at the expense of the coastal zone fishery, which "has been reduced to a part-time, seasonal and increasingly specialized activity" (88). Government promotion of a more specialized, capital-intensive fishery was also identified as a primary cause. In this presentation, the coastal zone fishery was depicted as struggling to survive, but the draggers were also in crisis because their technology resulted in overfishing the stocks on which they depended.[2] In this particular area, it would appear that small-scale capital, especially the DCP coastal zone fishers, was seriously threatened. "Few indicators hold out much hope for the immediate future of this particular fishery" (Davis 1991:93). This image contrasts with Apostle and Barrett's (1992) positive, if not unproblematic, vision.

These views of the fishery in Nova Scotia mirror the broader literature on DCP in which, for the past twenty years, two contrasting positions on the fate or trajectory of DCP have been projected. In the one case, capitalist development is associated with the eventual elimination of DCP whereas, in the other, persistence and even expansion is to be expected in certain types of economic activity (summarized in many sources including Sinclair 1985; Reinhardt and Barlett 1989; and Eder 1993). In recent years, a position is seldom held without qualification and it is common to find warnings against simplistic generalizations. Referring to agriculture, Eder (1993:647–48) usefully divides the factors that contribute to persistence into "countervailing contextual circumstances," which include supportive state policies, the technical process of production, close urban markets, and off-farm employment, and the "characteristics of the farmers themselves," which include the organization and orientation of the household to meeting minimum consumption needs rather than to making profit. To some degree these factors are said to provide insulation against capitalist appropriation or even to make DCP competitive (Reinhardt and Barlett 1989:204–05). Like many recent authors, Reinhardt and Barlett are equivocal about the trend; having defined the family farm in a way consistent with our understanding of DCP, they go on to note that the types of conditions mentioned

> can swing the balance in favour of one or another organizational form. The interaction of these factors can generate a complex pattern of agrarian structure, with family farms successfully competing in some spheres of agricultural production and capitalist units predominating in others (1989:207).

Applied to DCP in general, we share this perspective with its implication that outcomes are contingent on the circumstances that prevail in particular contexts, and thus that generalization is risky. Consideration of changes in the last decade in western Newfoundland, however, leads to the conclusion that the share or space for DCP in the fishery of this region is much more restricted than Sinclair (1985) had previously believed.

Before the extent of the influence of these factors on the fate of DCP in northwest Newfoundland can be understood, it is necessary to set the stage. Hence, a brief description of the geography and history of the area will be presented.

The Setting: The West Coast of the Great Northern Peninsula

The Great Northern Peninsula juts north roughly 300 kilometres towards Labrador from the heart of the island of Newfoundland and separates the Atlantic Ocean from the Gulf of St. Lawrence to the west (see Map 1). The Long Range Mountains dominate the southern two-thirds of the GNP, almost as far north as Plum Point. They form a plateau at 600–700 metres, reaching their maximum elevation of 806 metres on the summit of Gros Morne, which overlooks the northern shore of Bonne Bay. Along much of the west coast we find a coastal plain separating mountains from the sea, whereas on the east coast the plateau dips precipitously into the ocean. At St. John Bay, the western mountains touch the sea once more, but north of Plum Point they fade into a flat, marshy plain before re-emerging as barren hills on the northern tip of the peninsula. With the few exceptions that have been noted, the western coastline

Map 1. The Great Northern Peninsula and Surroundings

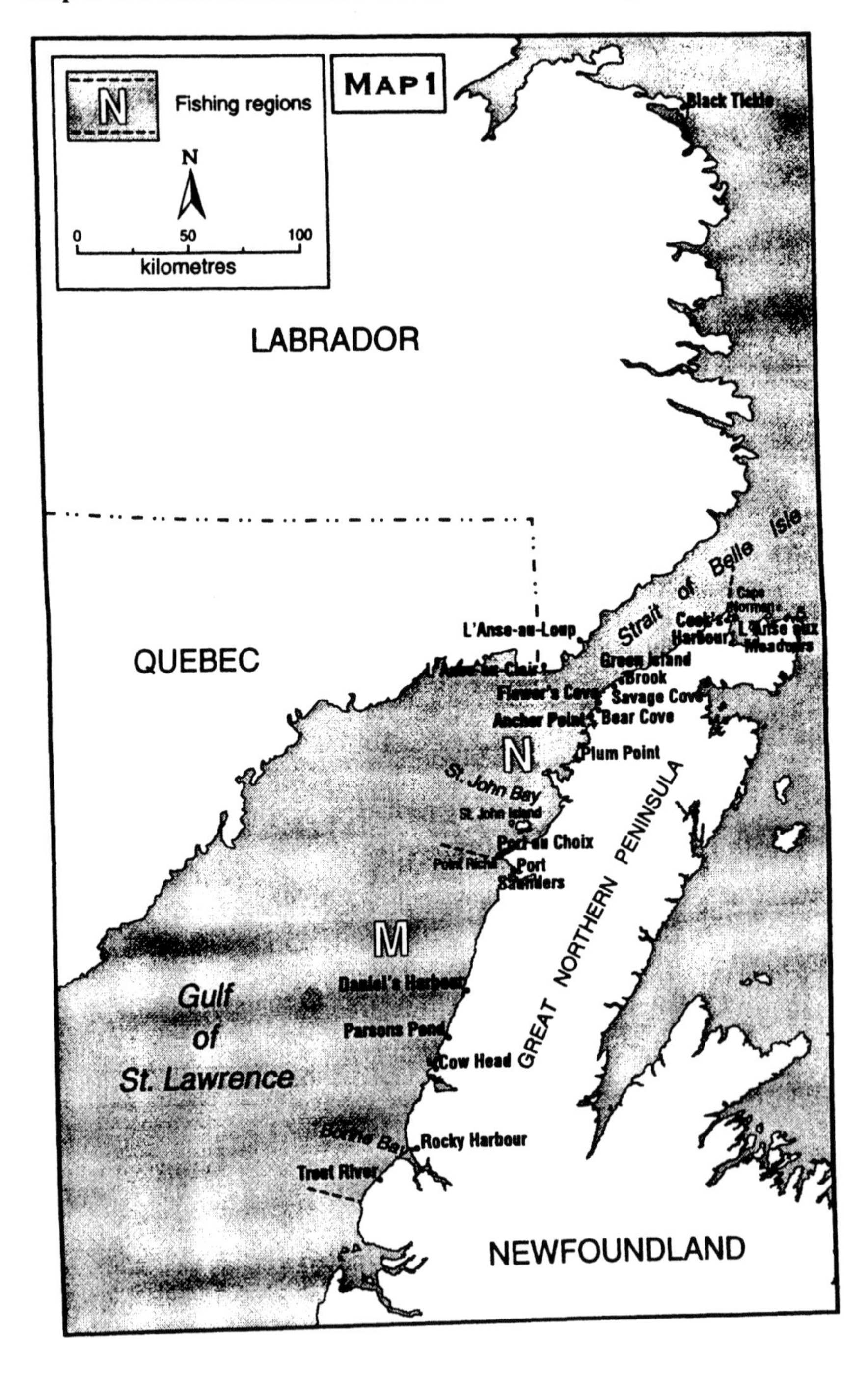

is generally low, rocky and windswept. North of Bonne Bay, few natural harbours are present on the west coast; Port Saunders, a deep inlet protected from open-sea storms is the safest.

The climate of the west coast of the GNP deteriorates as one moves north towards the Strait of Belle Isle, which has a near arctic climate (Hare 1952:46). A major factor contributing to the severity of the climate at the northern tip of the peninsula is the cold Labrador current, which splits in its southward journey and travels along both sides of the GNP. This water mass generates considerable fog and brings pack ice into the area most winters, where it adds to the shore-fast ice that forms along the northern coast of the GNP. This accumulation of ice generally closes the harbour at Port au Choix from mid-January to early April (Black 1966:26) and may linger in the Strait of Belle Isle well into July (Hare 1952:46).

The short growing season, resulting from cold temperatures and the poor soils created by past glaciation, means that the GNP offers very little possibility for agricultural development. Although forests gave rise to a significant timber industry in the past, the importance of this resource has dwindled in recent years.

Table 1. Home Communities of Dragger Skippers Active as of May 1994, Great Northern Peninsula and Southern Labrador

Community	*No. of Active Dragger Skippers*
Port au Choix	19
Anchor Point	14
Port Saunders	8
Savage Cove	8
Black Duck Cove	5
Bear Cove	3
Cook's Harbour	3
L'Anse au Loup (Labrador)	3
L'Anse au Clair (Labrador)	2
Sandy Cove	2
Bartlett's Harbour	1
Blue Cove	1
Brig Bay	1
Castor River North	1
Eddie's Cove West	1
Green Island Brook	1
Reef's Harbour	1
Shoal Cove West	1
St. Anthony	1
Total	76

Source: Regional provincial fisheries.

Map 2. Fishing Areas around Newfoundland

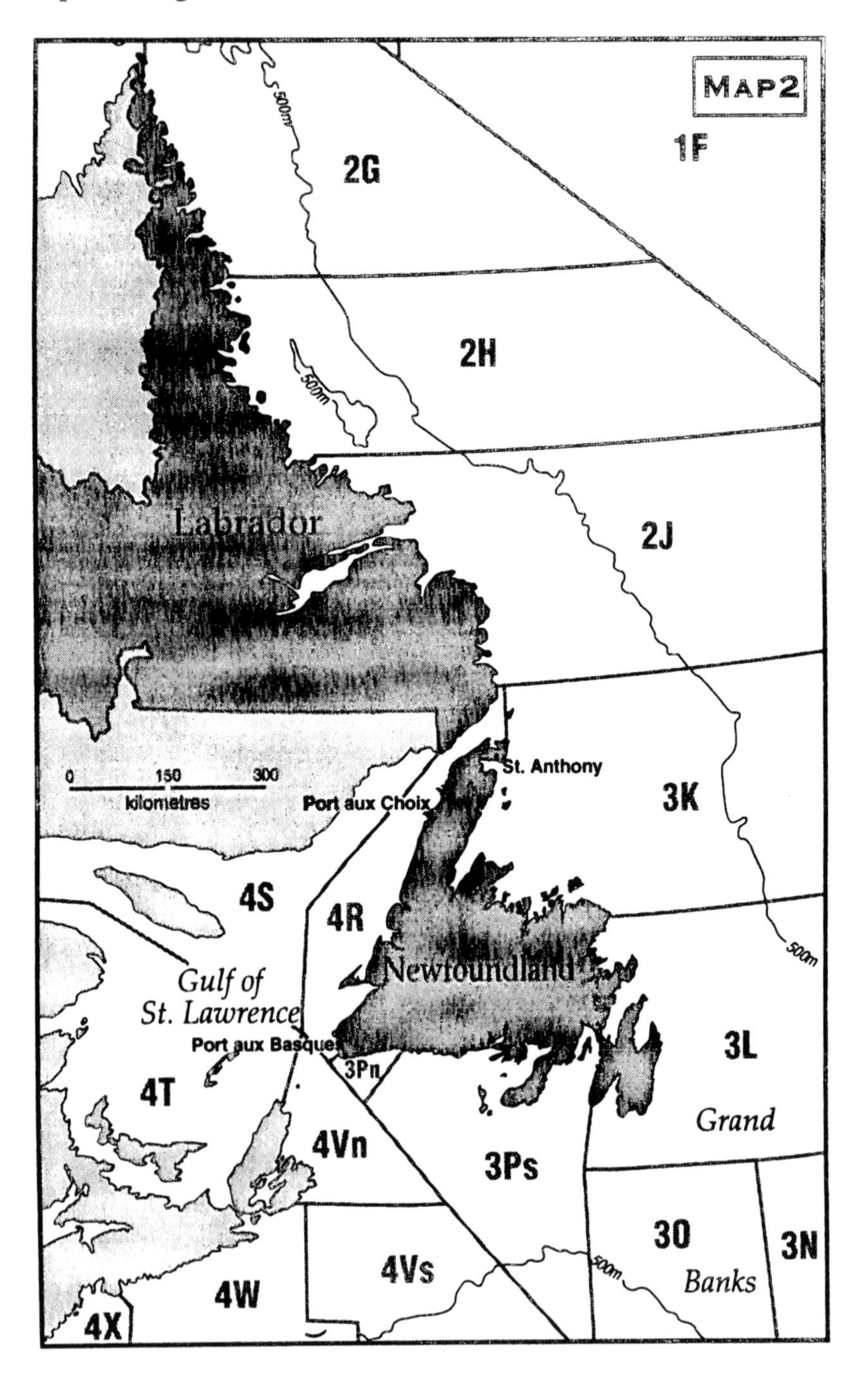

The old rocks of the Long Range Mountains may well contain important mineral deposits, and at least a small amount of petroleum is known to exist near Parson's Pond. However, the zinc mine at Daniel's Harbour is now closed permanently and the absence of other developed natural resources leaves the marine environment crucial for the area.

The northwest Newfoundland dragger fleet is located on the northern half of the western side of the GNP. Although dragger skippers and crew members live in communities from Daniel's Harbour in the south to Cook's Harbour and the Labrador side of the Strait of Belle Isle in the north, the fleet is clustered in Port au Choix and Anchor Point, with secondary centres in Port Saunders and Savage Cove (see Table 1). Due to its location on the western side of the GNP, the dragger fleet has depended almost exclusively throughout its history on fish from the Gulf of St. Lawrence in areas 4R and 3Pn and has been officially excluded from the fish stocks of areas 2J, 3K and 3L since 1982 (see Map 2). In addition to the groundfish species of cod, redfish, turbot, plaice, halibut and witch, the Gulf of St. Lawrence has provided herring, salmon, capelin, mackerel, lobster, shrimp, scallops and seals. These resources have been the key to human settlement in the area.

History of Settlement

Dorset Eskimo and Maritime Archaic Indians are known to have inhabited the Point Rich Peninsula thousands of years ago; and, about 1000 AD, L'Anse aux Meadows became the site of the first European settlement in North America. Archeologists have found traces of Vikings in this cove, which has been designated a World Heritage site. Much later, in the sixteenth century, Basque and Breton fishers caught whales and cod from shore stations in southern Labrador. They were followed, after the defeat of the French in Quebec, by English migratory fishers.

Settlement of the Newfoundland shore, however, is of more recent origin. Thornton's (1977:164) research records only five settlers in the region prior to 1825, all of them from England. Reporting local folk accounts, Canon J.T. Richards (1953) placed the first settler at Anchor Point no later than 1750. After more than 100 years of settlement, if 1750 accurately dates the beginning, the population of the entire northwest coast, from Cape Norman to Bonne Bay, amounted to only 490 (Newfoundland 1857).

This slow pace of growth can be attributed to several causes, but perhaps the most crucial factor was the complex relationship between French and British rights to this area. The French Shore treaties are extremely significant to the history of settlement in the region. The term "French Shore" refers to the portion of the island's coast on which French fishers were entitled to dry their catch. In signing the Treaty of Utrecht (1713), France no longer claimed ownership of Newfoundland but received the right to process fish from Cape Bonavista on the northeast coast, round the tip of the Great Northern Peninsula and down the west coast as far as Point Riche. In 1763, Britain ceded St. Pierre and Miquelon to France to provide a south coast base for French fishers. In the nineteenth century, this was to become the bustling centre of a large French banks fishery and the source of much resentment on the part of Newfoundland merchants. In 1783 the boundaries of the French Shore were altered by the Treaty of Versailles, by which the British retained more of the northeast coast and conceded the whole western shore to the French. Thus from 1783 until the final elimination of French rights in 1904, the French Shore extended from Cape St. John to Cape Ray. Throughout this time Britain accepted the French entitlement to use the shore without hindrance from British or Newfoundland fishers but did not accede to the French claim to exclusive fishing rights. Yet Britain discouraged settlement, since settlement would no doubt have provoked conflict with the French, and was extremely slow to recognize the rights of those who did choose to make homes on the French Shore (Thompson 1961; Neary 1980).

Thus French fishers were not permitted to settle in northwest Newfoundland and British fishers were discouraged from doing so. However, several British settlers did establish themselves and maintained reasonably peaceful relationships with the French, often by looking after their gear over the winter. Settlement was extremely limited until well into the 1800s, however, due largely to a shortage of women (Thornton 1977:166–67). This occurred because the men who became the first permanent residents were mostly from southwest England and Jersey rather than from other parts of Newfoundland and had arrived as young single assistants to British merchants on the Labrador shore. Thornton observes that the rate of permanent settlement speeded up only when the nine daughters of one of the region's couples reached the age of marriage. These newly married couples then settled in different coves along the coast, thereby founding many of the communities still in existence today (see Firestone 1967). The practice of patrilocality has resulted in a high percentage of the current residents of these communities having the same last names as the original male settlers.

The Inshore Fisheries, 1850–1965

The year-round English-speaking inhabitants appear to have been tolerated by the French, on what was still the "French Shore" until 1904, as long as they did not interfere with French fishing practices, which consisted of a seasonal longline fishery for cod centred in Port au Choix. Lobstering became a commercial fishery in 1873 when a Nova Scotian merchant established a lobster canning factory at St. Barbe. Other factories followed and, by 1888, twenty-five factories, employing about 800 local people, were operating on the west coast. Fishers even began abandoning the cod fishery in preference for lobsters, which provided a higher and more secure income (Campbell 1888).

Throughout the period 1905–65, the northwest coast fishery was conducted within the organizational framework of DCP. With few exceptions, the fishing crews were composed of kin who shared expenses and income based on the volume of cured fish that they delivered to their local merchant at the end of the fishing season. The women and children who helped process the catch were not directly compensated. Payment of a full share of the boat's income implicitly recognized that the person concerned contributed both capital and labour. This is evident in the treatment of the occasional outside (i.e., unrelated) shareman, who worked for a half share, plus room and board (Sinclair 1985:43–45).

Judging from data available for the census years 1911, 1921, 1935 and 1945, the total number of persons involved in the fisheries fluctuated at around 1,500 throughout most of the period. In 1945 only 3.7 percent of fishers were identified as wage-workers, which confirms the DCP nature of the fishery of this period (Sinclair 1985:43). From 1956 to 1965, the federal government's statistics show considerable yearly variation, reaching a peak of 1,433 fishers in 1959 (Sinclair 1985:38). There were always more fishers along the northern coastline (area N on Map 1), where fishing was also more likely to be the only reported occupation. South of Point Riche (area M on Map 1), many combined their fishing season with woods work or construction. In the northern area, cod was supplemented by salmon and herring fisheries. South of Flower's Cove, lobsters continued to be an important species during the period of consideration, providing an additional source of income that was not available to the northern fishers. The latter, however, were better placed to prosecute the cod fishery. In a good season during the 1940s, about 25 quintals of dried fish were produced per fisher (calculated from data in Newfoundland Fisheries Board 1940–48). This fish was sold at prices varying from approximately $7.50 per quintal for small West India grade to $15 for large and medium merchantable. Even if all fish were top quality, which is extremely unlikely, the average gross revenue per person could not have exceeded $375.

The lobster fishery was fairly stable throughout the period until 1957, with 65,000–70,000 traps in use; traps then increased to 172,000 by 1965. Lobster was harvested in small two-man boats during the time the shore was free of ice (Sinclair 1985; Palmer 1994). Around 1950, the typical lobster fisher, operating alone or together with another fisher, was setting about 100–150 traps and harvesting about 1,700 pounds of lobster per year.

The variety of production techniques in the cod fishery is impressive, as is the apparent readiness of fishers to adopt new technologies. One common fishing technology was the baited longline, which typically consisted of about fifty hooks placed along 50 fathoms of line (one fathom equals 6 feet, slightly less than 2 metres). These would usually be set in fleets of six to eight joined lines. Each end would be secured by a grapnel and marked by a buoy. Another common device was the handline, which consisted simply of a single fishing line with several baited hooks. Another method of fishing was jigging, in which the fish is attracted by a lead lure to which a hook is appended; the fisher then jerks the line suddenly and, with luck, the cod is caught. Cod nets, which were moored in berths to trap fish by the gills, were also used but were not nearly as effective as the modern monofilament gill nets (described in Chapter 2) that came into use at the end of the period under consideration.

Despite this variety of fishing techniques, it was the cod trap that largely determined the technological and social structure of the northwest coast. For those who were prepared to carry a debt with the local merchant that was considerably larger than normal or, in rarer cases, for those who had accumulated sufficient capital, the cod trap was the most important component in their fishing gear. The cod trap is a type of fixed gear that is essentially a large box of net, from 47–60 fathoms around its perimeter, with an opening, or "door," on one side and a leader that directs fish into the opening. As stationary gear, it requires a berth, and the number of traps in use must be limited by the number of suitable berths. These berths are usually either owned by a family and inherited patrilineally, distributed within a community on the basis of an annual draw or rotated among different families on a yearly basis. Invented in 1871, the cod trap was adopted swiftly in northwest Newfoundland, probably from migratory Labrador fishers who fished from the island shore as well as across the Strait of Belle Isle. By 1911 there were 277 cod traps reported along the northwest coast. In 1921, 350 traps were moored in these waters, but subsequently this technology underwent a decline until, in 1956, only 87 were reported. Between 1957 and 1965, however, their number rebounded again to average 202 (Sinclair 1985:37–39).

Cod strike inshore, where cod traps are moored 50–60 fathoms from shore, from early to late June, later in the north than the south, and remain for three or four weeks. At this time the cod trap can be highly productive compared with the old cod seines or nets, although trap fish are smaller on the average than those caught by other gear in deeper water. If for any reason fish stay offshore, the season may prove disastrous for the trap fishery. In addition, the work of setting and hauling (or "tucking") the trap requires a team of four to six fishers and two boats. Thus, the trap fishery requires more labour and capital than other inshore techniques.

Although most fish merchants bought some fish fresh "from the knife" and subsequently dried or pickled it on their own premises, the majority of fish was salted by the fishers and their families. Although a few pickling plants purchased cod in salt bulk form, sun-dried fish was the most common method of processing. After being left in salt bulk for eight to ten days, the split fish was washed in seawater and then dried in the sun on flakes or spread out on the beach, the total process taking about thirty days, depending on the number of good curing days (Proskie 1951:22). The labour time required to produce the cure inevitably reduced fishing time in the summer and prevented a late fall fishery because curing was usually impossible after the beginning of November (Proskie 1951:44). Whole families—men, women and children—participated in the curing process, with women being particularly involved in drying. According to Found (1963), the typical salt-cod fisher spent 30 percent of his time processing the catch.

Subsistence production and occasional wage labour, in conjunction with transfer payments after confederation with Canada in 1949, helped DCP producers to survive, but life was difficult and migration was a common response (Sinclair 1985:56). In summary:

> The overall consequence of this system was that only the exceptionally fortunate crew could produce enough fish to accumulate any capital over the fishing season. Usually, the fishermen began the season on debt and ended it with a modest surplus, if any at all (Sinclair 1985:46).

Notes

1. A condensed version of the monograph appears as a chapter in Apostle and Barrett (1992).
2. Davis identifies the same kind of social and cultural divisions between coastal zone and dragger fishers as Sinclair (1983, 1985) found in northwest Newfoundland.

Redfish dragging

Collector boat at Black Tickle, Labrador

2
The Rise of the Dragger Fleet, 1965–82

The thirty years after 1965 were a period of innovation and change in the fishery and other aspects of Newfoundland society. Great Northern Peninsula fishers experimented with new fishing technologies and added shrimp and scallop to the cod, lobster, herring and salmon they had long fished. Since new technology like the otter trawl was costly and required larger boats in order to fish further from home, these innovations altered, at least for some, the DCP character of their fishery. By 1982, Sinclair (1985) concluded that some fishers were operating small capitalist enterprises. However, it was also obvious that these changes had produced problems in the relationships between fishers and processors, fishers and government, and even among fishers themselves. Less apparent in 1982 was the ecological danger produced by these changes. Hence, although the following analysis of the rise of the dragger fleet between 1965 and 1982 will draw heavily on Sinclair's earlier work (1985), it will be augmented by observations about ecological trends that take on new significance when viewed from the perspective of the 1990s.

The Introduction of Longliners

Although longliners[1] appeared on the southwest coast and in the Bonavista area in the early 1950s, this type of vessel was not introduced to the Port au Choix area until 1963. The tardiness of this innovation is due to a number of factors. First, the system of DCP and merchant capitalism produced little or no surplus for fishers to apply in building up a longliner fishery. But, this was a widespread problem in Newfoundland and is thus insufficient in itself to account for the delay. Second, the isolation and inferior education of most people in the area had created a set of low expectations that restricted the prospects of acting to improve living conditions. Where greater expectations existed, plans were often thwarted by the tradition of cooperative family enterprise, which made it difficult for the more ambitious individuals to withdraw from the family crew or persuade the family as a whole to risk a new kind of fishery. In 1963, however, spurred on by a desire for a better standard of living and by knowledge that the stocks fished from offshore trawlers in the Gulf might be accessible to longliners, several Port au Choix fishers built homemade longliners. This was the start of a process that was to change the nature of local fishing and transform Port au Choix from a small village of part-time lobster fishers (there were no full-time fishers recorded in 1963 [Black 1966:81]) into the single most important fishing and processing harbour on the coast.

The number of longliners soon increased along the coast after the initial vessels proved a success. In 1965, only eight to ten longliners were delivering to the fish plant in Port au Choix, but this number increased quickly to seventeen in 1966 and forty-eight by 1968. During this period most fishers substituted gill nets for longlines, which required much tedious labour because each hook had to be individually baited. The gill net is essentially a mesh fence constructed of thin, strong nylon and measures approximately 50 fathoms long and 25 meshes deep. It is weighted at the bottom and settles on the seabed (Brothers 1975:29). Although gill nets improved the situation for longliner fishers—relative to their initial reliance on trawl lines—this technique was rather quickly superseded by dragging with an otter trawl. Gill-net longliners did not disappear, but they came to occupy a secondary place to the nearshore draggers, because most gill-net fishers converted their longliners to draggers during the early 1970s.

From Gill Nets to Draggers

In light of recent events in the fishery, it is ironic that overfishing, along with high costs relative to income, was the chief cause of the switch from gill nets to small draggers. The gill netters began to struggle in the mid-1960s when large trawlers, both foreign and domestic, along with factory ships increased fishing in the Gulf of St. Lawrence. We shall reserve the term *draggers* for vessels under 65 feet that use otter trawls and shall refer to vessels over 65 feet as *trawlers*. Although there was some concern about the health of the stocks at this time, the main problem was perceived to be the destruction of gill nets by trawlers, including trawlers illegally fishing inside the three-mile limit. By the late 1960s, however, the decline in the cod stock due to overfishing had become obvious to local fishers. In response to this situation, and a period of low prices for fish, gill netters pressured the federal government to ban the draggers from inshore grounds. At the same time, they also increased the numbers of nets they used from about ten to forty per vessel. This strategy, in addition to aggravating the overfishing problem, also increased the gill netters' investment cost from about $900 to $3,600. As a result, many gill netters began to search for new technologies as a solution to their dilemma.

The first step in the move away from gill nets was dragging for scallops south of Port au Choix in 1969. Although scalloping never became a significant fishery until new beds were discovered between Cook's Harbour and Eddie's Cove East in 1984, the move away from gill netting had begun. The transition accelerated when ten Port au Choix skippers rigged up their longliners with dragging gear and started the shrimp fishery in 1970 in the Esquiman Channel, 20 kilometres west of Point Riche. Initially, the fishers had no more than modest success because their boats were small and underpowered, with engines generating from 70 to 80 h.p. compared with a maximum of 625 h.p. in 1982 and 850 h.p. in 1992. In those days, too, the net twine was less durable than at present, the shape of the nets was less efficient and "tear ups" were more common. Nevertheless, the gear cost no more than a fleet of gill nets and it was possible to make a better living. Hence, new entrants were quickly attracted.

From the beginning, and continuing through the 1970s, Canada's federal government was closely connected to the rise of the new fishery. It created the conditions that allowed local skippers to flourish if they could marshall sufficient initiative and resources. Exploratory research on the shrimp stock was followed by controlled access through licencing and substantial subsidies to permit the acquisition of vessels with sufficient power to drag efficiently for shrimp and cod. Licencing policy was particularly important because it solidified the emerging social structure in such a way that the new elite status of dragger skipper could not be penetrated by others.

By 1975 the shrimp fleet was attracting an increasing number of new entrants, and skippers began to fear overcrowding. They requested that the federal government restrict entry to the existing licenced operators (Fisher 1977:18). Given the preference of state officials for such a policy, limited entry licencing was introduced for shrimp fishing beginning in the 1976 season when there were thirty-nine operating vessels with another three applications awaiting a decision. The latter were approved. However, the poor season of 1977 brought renewed protest from experienced shrimp fishers that their livelihood was threatened by new entrants. In an attempt to reduce the fleet, the federal government initiated a utilization clause attached to the licence, whereby the operator had to land at least 70,000 pounds of shrimp in 1978 in order to retain his licence. The decline in the number of licences issued for 1979 represents the application of this clause (Fisher et al. 1980:2). With the rise of a directed groundfish dragger fishery, officials with the Department of Fisheries and Oceans (DFO) reinstated a limit of forty-two shrimp licences, a limit that was to increase to forty-nine in 1986 and fifty-seven in 1991 (Cashin 1993).

The development of the shrimp fleet was assisted by generous federal and provincial financial support to upgrade vessels. In 1976, thirty of the thirty-nine boats in the fleet were 52–58 feet long. None was under 50 feet, and the first 65-foot vessel had been purchased in the previous year (Fisher 1977:2). This number would gradually increase over the next few years, with two $850,000 steel-hulled draggers entering the fleet in 1983. Without public subsidies for vessel construction, it is unlikely that fishers would have been able to develop the shrimp fishery so quickly, for the returns from the fishery itself would not have permitted the necessary level of savings.

Although the shrimp fishery is highly variable in terms of yearly fluctuations and variations between boats, it clearly reached a peak in 1979 as measured by volume of landings. The subsequent decline did not result from a scarcity of shrimp but rather from the expansion of cod dragging, which provided higher returns for less effort. As participants remember it, several fishers started to chase cod seriously in 1974–75—"We tried catching cod with our shrimp nets," but without much success at first, because competition from larger trawlers, lack of electronic gear (especially a depth sounder to locate fish and provide information on the state of the bottom) and inadequate power detracted from their efforts. Furthermore, small, underpowered vessels could hardly drag a large cod net, whereas, moving with the tide, it was possible to handle shrimp fishing. Of course, cod dragging continued to be carried on from the older smaller boats, but they were incapable of landing anything approaching the volume of the larger 65-foot boats. As it became evident that larger boats could work in worse conditions and catch more fish, skippers were increasingly anxious to invest in new vessels and opt for cod dragging on a large scale.

For many, the turning point was their experience in the "winter fishery" at Port aux Basques (see Map 2). Significantly, residents of Port aux Basques and other harbours on the southwest coast of Newfoundland had long practised a winter cod fishery, but this resource was beyond the needs and capabilities of Great Northern Peninsula fishers until the development of the dragger fleet. In 1977, fifteen skippers from the Port au Choix area decided to venture south to try their luck at the winter fishery: "People thought we were crazy and still call us the Vikings from the north." Braving the harsh winter conditions, the northern fishers searched out areas unfished by locals and discovered "more fish than we'd ever seen before." But only one boat was successful—a "65-footer" that had the power to drag in these conditions caught 300,000 pounds, whereas the others had to return home early. Their reaction was to acquire technology to capture the fish. "It was the possibility of catching all that fish at Port aux Basques that got us into the big boats."

By 1982, the winter fishery was prosecuted with great success by nearly fifty draggers. The largest landing recorded in a single day (according to local knowledge) was 120,000 pounds, and up to 350,000 pounds were taken in a single week. With catches like this, it is not surprising that the quota, set by the federal government since 1977 on each major stock and fleet sector, would be exhausted after four to five weeks of fishing. However, the large number of vessels engaged in the fishery in 1982, and the imposition of weekly quotas by local fish plants to avoid gluts, reduced the total catch that a skipper might expect to about 300,000 pounds. Still, several skippers were so successful at cod dragging that they reported over two million pounds of groundfish landings per year during the early 1980s. The new dragging technology made it easier to make money at cod than at shrimp.

As cod dragging became visibly profitable in the late 1970s, this technology became increasingly attractive to those ambitious fishers who were restricted by their licences to gill-net operations. Therefore, these fishers pressured the federal licencing authority to issue additional permits. With few exceptions, it was now fishers from outside of Port au Choix and Port Saunders who wanted to be able to participate in the fishery. At first, the federal fisheries department held firmly to its policy of refusing any additional licences, but in 1979 approval was given for six otter-trawl licences for groundfish operations.

Fishers who purchased longliners after 1976 were probably well aware that they could not obtain otter-trawl licences but hoped that the spectre of large vessels unable to catch enough fish would force the DFO to issue new entry permits. If this was indeed the strategy, it was partially successful. However, the six new licence holders were limited to 700,000 pounds each (a restriction soon lifted) and could not take part in the winter fishery. At the time, this policy enraged other local fishers who had been denied this preferential treatment and failed to satisfy the new recipients, who felt that they should be restricted only by the general fleet quota. By 1981 the situation escalated into a mass protest against the licencing policy, with the consequence that Fisheries Minister Romeo LeBlanc decided to open the cod-dragging fishery to an additional forty-nine vessels. By August 1982, thirty-two fishers had taken up their option, with the remaining openings going unclaimed because some longliner skippers were unwilling to take the financial risk of investing in the larger boats necessary to make dragging profitable. The quota for groundfish otter trawlers was separated from groundfish allocations to the shrimp fleet, possibly to encourage the latter to devote more attention to shrimp by removing the fear that their cod allocation would disappear into the holds of these new entrants.

The Fishery in 1982

As a result of the events just described, the northwest Newfoundland fishery had become exceedingly complex by 1982. In addition to those fishers who adopted the dragger technology, many continued to operate small boats and to fish lobster, salmon and cod, while others persisted with a gill-net longliner operation. Indeed, scattered along the coast from Trout River in the south to Cook's Harbour in the north, 2,917 residents held fishing licences and almost exactly half of these worked full-time at the fishery (calculated from data supplied by the DFO, St. John's). The nine communities which had at least fifty full-time fishers contained only 45 percent of all full-time fishers but included most of those who had switched to capital-intensive technologies.

Lobster and salmon licences, sometimes held in combination, were extremely important, particularly on the southern part of the coast. Holders of scallop licences in the Port au Choix area had ceased to use them, but this fishery continued to be important further north, especially on the Labrador coast. Shrimp and groundfish trawl licences were heavily concentrated in Port au Choix, Port Saunders and Anchor Point.

It is evident that, although the dragger fishery had been a highly visible and important new form of fishing, it had not absorbed the majority of fishers in the area. If we estimate an average of four crew members to each of seventy-five draggers, the total number engaged in the dragger fishery would be three hundred. Allowing three persons to man each of the other 101 longliners would make a labour force of 303, and a total of 603 on all vessels over 35 feet (calculated from data supplied by the DFO, St. John's). Although these figures are rough estimates, it is fair to conclude that only about 35–40 percent of *full-*

time fishers were engaged in this more capital-intensive fishery. Further, nearly all the part-time fishers worked on small boats. Although the dragger fishery dominated a few of the larger settlements, Port au Choix especially, it was not the most common form of fisheries production on the coast; indeed, the dragger skippers constituted a local elite in contrast with most struggling small-boat operators. The material basis of this social division is evidenced by the fact that in 1981 the mobile-gear catches of groundfish exceeded that of the fixed-gear fleet (Sinclair 1985:80), and the value of shrimp, scallop and groundfish landings by mobile-gear vessels was 15 percent greater than the total value of all species caught by other techniques (derived from Sinclair 1985:81).

By 1982, northwest coast fishers practised one of two strategies. The majority fished using the "traditional" methods of DCP, while a second, newer group of dragger fishers used new technology such as otter trawls to fish further from home. Dragger skippers, and to a lesser extent their crew, were separated from other fishers not only because they had become visibly more affluent, but also because they appeared to have shed most of the attributes of DCP. Rather than own and operate vessels with other family members in order to meet the basic needs of their households, most skippers appeared to have become sole owners who hired (often non-kin) labour and vigourously pursued profit. In other words, they functioned as small-scale capitalists. In his previous work, Sinclair (1985) recognized that this transition was ambiguous and incomplete. In Chapter 6 we will return to the important consideration of the extent to which dragger fishing marks a transition in social organization away from DCP towards capitalism.

Despite the success of the dragger fleet as a whole in 1982, there were reasons for some of the skippers to worry about the future. Many new entrants (approximately 50 percent, according to a senior fisheries official) had purchased vessels at $400,000 or more. For these fishers, and for many of the established skippers, the burden of mortgage payments as high as $70,000–$80,000 per year threatened bankruptcies and a rash of insurance claims. Hence, an expanding supply of groundfish in the Gulf of St. Lawrence was crucial to the future of the dragger fleet. Although government projections predicted neither a major decline nor an increase in the stock itself (*Northern Pen,* February 11, 1982), many dragger fishers were optimistic. It was thought that the end of France's access to the Gulf stock in 1986 would provide a major increase in fish available to the dragger fleet. Further, many fishers reported cod to be more plentiful than in the previous few years and it was generally expected that their quotas would increase.

Conclusion

In 1982 the majority of the population of northwest Newfoundland was still engaged in DCP. The social landscape and the structure of the fishery had been irreversibly changed, however, by a small minority of entrepreneurial fishers who, with the aid of supportive state policies, had adopted a more capitalistic form of production. Questions began to crystallize concerning the future of these trends. Would the dragger fleet continue along its path away from DCP and rely on a greater use of non-kin labour and the adoption of an expanded mode of reproduction? Would the number of fishers using draggers remain stable or would there be further protests for more licences? And, perhaps most importantly, what would happen to the majority of fishers still engaged in DCP? Could they co-exist economically and socially with their capitalistic neighbours, or would the rise of the dragger fleet spell the end to DCP in northwest Newfoundland? From the perspective of the mid-1990s, the important question being ignored in 1982 concerned the impact of these technological changes on the ecology of the Gulf cod stock.

The relative lack of ecological concern in 1982 is particularly interesting because the causes of the technological differentiation and expansion that make up the story of the northwest Newfoundland fishery from 1965 to 1982 reveal a paradox. On the one hand, these changes were motivated by a desire, on behalf of both fishers and the government, to increase the amount of fish caught through technological modernization. On the other hand, most of the significant steps in this transition were actually the results of overfishing and a realization that fish stocks were being threatened. This category of events includes the shift from gill nets to draggers and the subsequent limitations placed on the licences and quotas of the dragger fleet. The threat these changes presented to the cod stocks were never fully appreciated, however, and the brakes placed on the dragger fleet failed to offset the increases in harvesting capacity brought on by increases in skill and technological advances. What awareness there was about ecological concerns could not overcome the "optimism" placed in technological advancement, especially when associated with short-term economic improvement. It would take another decade for this situation to change.

Note

1. In Newfoundland the term *longliner* is commonly applied to decked vessels between 35 and 65 feet long, regardless of the kind of gear used to catch fish. Strictly speaking, the name is appropriate to describe the early boats which were built to extend the season and the range for fishers working with trawl lines, but the label has persisted, leading to some confusion among those unfamiliar with Newfoundland fishing categories. Along the northwest coast of Newfoundland the term *dragger* is normally used to refer to the fleet of small, net-dragging vessels, and *longliner* is reserved for other decked vessels, which are usually shorter.

Scallop longliners

Icebreaker opening up Port Saunders for dragging in the spring

3
The Glory Years, 1982–87

While the 1970s had seen the emergence of the dragger fleet, the 1980s would be concerned with the challenge of managing this fleet. Despite the limited entry licencing introduced in 1976, the early 1980s remained years of expansion. Dragger skippers still strove to move into larger vessels with the consequence that the number of 65-footers grew from one in 1977 to seventeen in 1982. The fleet itself expanded with the new entrants of 1982, and the number of vessels going to the winter fishery also increased from about one dozen in 1979 to twenty in 1980, and approximately forty-five in 1985. The major force behind all this growth was the tremendous catch. Twenty-five-thousand-pound[1] tows and 50,000–80,000-pound days were common in 1981 and several 65-footers caught over two million pounds that year. Gross incomes of more than $200,000 were also being achieved by a number of the larger boats. This same period, however, began to see a decline in fixed-gear catches.

The rapid expansion of the dragger fleet soon brought the realization that it was necessary to find new means of limiting its harvesting capacity. This, in turn, brought concerns over the fleet's economic viability. Hence the period from 1982 to 1987 was dominated by a variety of management tactics, some quite innovative, aimed at reconciling the tremendous fishing capacity of the dragger fleet with economic realities. These tactics revealed a trend of growing ambivalence in state policy towards both DCP and the small capitalist draggers.

Given its economic importance to the entire fishing fleet and its social importance to the area, our analysis will focus on the cod fishery, although we do recognize that the region's fishery is based on many species. From 1981 to 1990, cod made up 59 percent of the northwest coast landings by value, followed by lobster at 15 percent, shrimp at 11 percent, scallops at 3 percent, salmon at 3 percent, crab at 2 percent, capelin at 2 percent, and a variety of other species making up the remaining 4 percent. Further, many of these other species were alternatives to cod, and the fluctuations in the other species during this time were largely determined by the ups and downs in the cod fishery.

The commercial importance of various species to fishers is best indicated by landed value, which is presented for the six most important species in the fixed-gear and mobile-gear sectors, with adjustments for changes in the value of the dollar, in Tables 2 and 3, and Figures 1 and 2. (Landed values in each year were altered to reflect the value of the dollar in 1986 as indicated by the consumer price index.) From 1982 to 1987, lobster and cod were the main species caught by fixed-gear fishers, with lobster coming to dominate after 1988 because of the decline in fixed-gear cod catches (discussed in Chapter 5). For the mobile-gear fleet, shrimp were consistently important, and in some years scallop also provided a substantial share of its revenues. Cod, however, was always the most important species and accounted for more than half the catch during the years 1982 to 1987. Cod dominated the social as well as the economic structure of the fishery because it could be taken by all licenced vessels, whereas shrimp and scallop could only be fished by trawl or dredge, and lobster licences were restricted to inshore boats. Hence, it was in the cod fishery that a major conflict between fixed- and mobile-gear fishers began to develop.

Table 4 presents information on the volume and value of cod landings by fleet sector. According to both measures, the fixed-gear fishery was experiencing a significant decline, despite a brief resurgence in value due to exceptionally high prices in 1987. The mobile-gear pattern is similar, but the relative share of both sectors in the cod fishery fluctuates. It is important to note that the landings on the northwest coast necessarily exclude the cod taken by mobile-gear boats from this area while fishing off the southwest coast in winter. Initially, the downward trend triggered economic rather than ecological concerns for most participants and managers, despite government estimates of a decline of 62 percent in the Gulf cod stock biomass, from a high of 432,000 tonnes in 1983 to 164,000

Table 2. Fixed-gear Landings, Northwest Newfoundland and Southern Labrador, 1984–92, Metric Tons by Rank Order of Value in 1992 (left=highest)

Year	*Lobster*	*Cod*	*Turbot*	*Capelin*	*Plaice*	*Herring*	*Salmon*
1984	711	12,883	45	227	440	1,560	54
1985	840	10,438	50	230	250	504	45
1986	723	7,367	27	113	176	931	77
1987	601	6,950	56	23	255	1,160	134
1988	652	5,862	122	996	345	764	96
1989	985	3,894	299	1,467	324	391	66
1990	784	2,938	159	1,640	209	646	40
1991	792	5,608	110	37	239	535	53
1992	861	4,769	475	1,089	203	609	16
Value of landings in constant 1986 dollars (thousands of dollars)							
1984	$4,064	$4,553	$15	$94	$146	$342	$198
1985	$5,475	$3,670	$17	$32	$81	$74	$175
1986	$4,214	$3,104	$10	$24	$67	$151	$289
1987	$3,878	$5,486	$43	$3	$140	$169	$489
1988	$3,219	$2,357	$42	$288	$177	$93	$357
1989	$4,665	$1,492	$169	$409	$151	$46	$227
1990	$2,664	$1,368	$126	$201	$93	$71	$130
1991	$2,780	$2,937	$262	$5	$104	$61	$170
1992	$3,901	$2,528	$362	$106	$87	$71	$62

Source: Department of Fisheries and Oceans, special tabulation.

Table 3. Mobile-gear Landings, Northwest Newfoundland and Southern Labrador, 1984–92, Metric Tons by Rank Order of Value in 1992 (left=highest)

Year	*Cod*	*Shrimp*	*Scallop*	*Mackerel*	*Capelin*	*Redfish*	*Herring*
1984	25,765	1,492	1,376	0	430	102	613
1985	14,296	1,291	2,149	0	21	146	55
1986	13,140	1,505	1,609	0	1,469	120	242
1987	12,214	2,287	973	100	661	153	498
1988	13,238	3,841	397	305	83	175	1,080
1989	8,566	2,989	109	402	1,211	170	961
1990	10,076	2,275	74	413	3	213	215
1991	6,570	3,686	400	1,155	65	275	545
1992	7,059	2,097	1,122	1,382	1,519	671	804
Value of landings in constant 1986 dollars (thousands of dollars)							
1984	$10,558	$2,158	$27	$0	$135	$27	$125
1985	$6,024	$1,684	$2,633	$0	$16	$38	$6
1986	$6,261	$1,998	$1,726	$0	$383	$36	$27
1987	$9,867	$3,890	$1,219	$17	$123	$7	$53
1988	$5,904	$6,182	$384	$35	$132	$46	$112
1989	$3,484	$3,597	$93	$56	$189	$37	$91
1990	$4,748	$2,107	$58	$56	$2	$49	$23
1991	$4,310	$3,444	$315	$196	$5	$57	$64
1992	$4,361	$1,611	$1,189	$178	$153	$129	$86

Source: Department of Fisheries and Oceans, special tabulation.

tonnes in 1988 (Frechet 1991). Only by the end of the decade did ecological issues become a major concern.

Management Tactics

In 1982 a new policy of sector management restricted the operation of the northwest Newfoundland dragger fleet to the Gulf of St. Lawrence and the southwest coast, while keeping the vast majority of larger trawlers (longer than 65 feet) out of these areas. The Gulf of St. Lawrence was, however, being fished by many draggers under 65 feet in length from Quebec, Nova Scotia and New Brunswick, and the stocks naturally migrated across fisheries sector management lines, which do not correspond to natural ecological units.

The northwest Newfoundland draggers had a good year in 1982 under the new sector management. Indeed, the main problem was severe glut conditions during both the winter fishery (February and March) and the summer fishery (May through July) as the harvesting capacity of the fleet was well above the processing capability of local plants. In addition to quality problems, such glut

Figure 1. Value of Fixed-gear Landings, 1982–92 in Constant 1986 Dollars

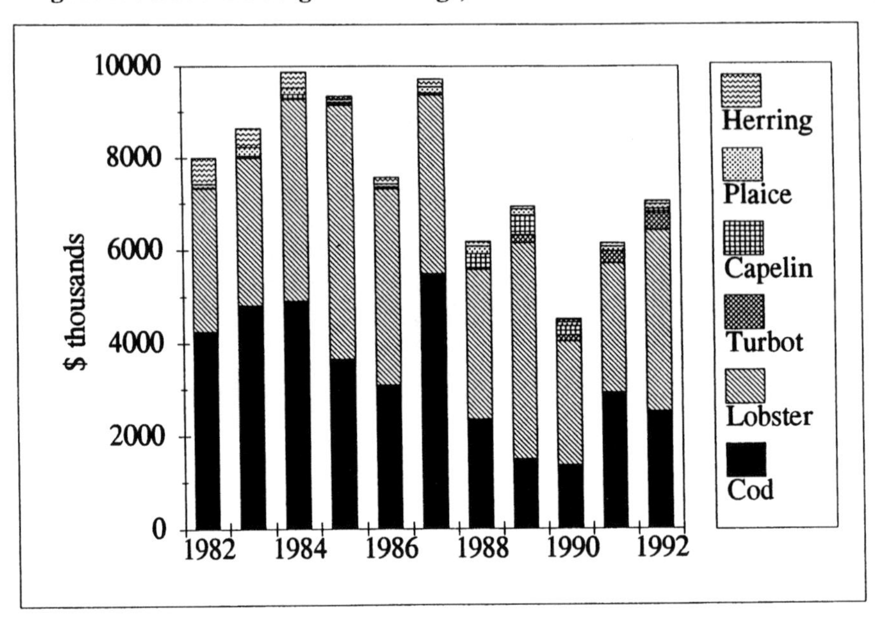

Figure 2. Value of Mobile-gear Landings, 1982–92 in Constant 1986 Dollars

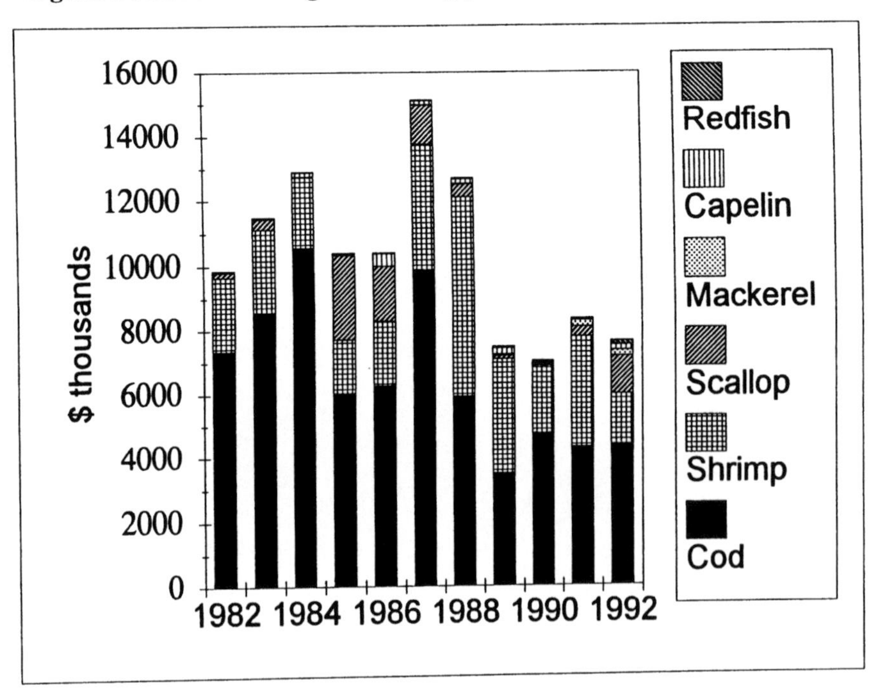

Table 4. Cod Landings, Northwest Newfoundland and Southern Labrador, 1984–92

Year	Fixed Gear	Fixed Gear as % of Total	Mobile Gear	Mobile Gear as % of Total	Total Landings
Volume of landings (metric tons)					
1984	12,883	33.3	25,765	66.7	38,648
1985	10,438	42.2	14,296	57.8	24,734
1986	7,367	35.9	13,140	64.1	20,507
1987	6,950	36.3	12,214	63.7	19,164
1988	5,862	30.7	13,238	69.3	19,100
1989	3,894	31.3	8,566	68.7	12,460
1990	2,938	22.6	10,076	77.4	13,014
1991	5,608	46.1	6,570	54.0	12,178
1992	4,769	40.3	7,059	59.7	11,828
Value of landings in 1986 dollars					
1984	$4,899,311	31.7	$10,557,837	68.3	$15,457,148
1985	$3,669,848	37.9	$6,023,369	62.1	$9,693,217
1986	$3,103,900	33.1	$6,260,768	66.9	$9,364,668
1987	$5,486,454	35.7	$9,866,534	64.3	$15,352,988
1988	$2,357,536	28.5	$5,903,349	71.5	$8,260,885
1989	$1,492,484	30.0	$3,484,280	70.0	$4,976,763
1990	$1,367,415	22.4	$4,747,840	77.6	$6,115,255
1991	$2,936,364	40.5	$4,310,251	59.5	$7,246,615
1992	$2,528,078	36.7	$4,360,397	63.3	$6,888,475

Source: Department of Fisheries and Oceans, Moncton.

conditions led to nearly ten million pounds of fish being trucked off the Great Northern Peninsula for processing elsewhere. In an attempt to avoid this situation, the fleet was placed on a weekly quota of 75,000 pounds, with buyers occasionally adding a further daily limit of 10,000–15,000 pounds. A crew member on a 55-foot dragger recalled these days fondly: "We never had any problem getting our [weekly] limit. In fact, we'd often go out Sunday night and be headed home [with the weekly limit caught] on Wednesday." Indeed, the fleet had caught their yearly quota of cod by the middle of July in 1982.

Although the daily and weekly limits slightly reduced the glut conditions, it also intensified concerns over the economic viability of many boats in the fleet. The larger draggers cost $600,000–$800,000 and would probably not remain viable at 15,000 pounds per day because such restrictions limited yearly gross incomes to slightly over $100,000. Further, the fleet now had more than thirty new vessels competing for a quota that had only increased slightly from the previous year. The economic pressure was also great on those boats—including all the new entrants—lacking a shrimp licence, because shrimp was worth between $60,000 and $75,000 in gross income per year. To ease hardship on non-shrimpers, the fleet quota of 25,000 tonnes of cod was split, with 10,000 tonnes going to boats with shrimp licences and 15,000 tonnes going to non-shrimpers. Still, the non-shrimpers caught their quota early and pressured the DFO unsuccessfully for more cod. They also pressed for more shrimp licences, and a few more were granted for 1983. In addition large by-catches of cod in the shrimp fishery began to attract attention, even though the cod was often given to other boats. The by-catches finally caused the shrimp fishery to be closed in November.

The weekly and daily limits of 1982 increased economic pressure on many boats without sufficiently solving the glut conditions and accompanying quality problems that had plagued the fishery. They also failed to eliminate the competitive nature of the fishery which was proving to be extremely dangerous, especially for skippers of smaller vessels who felt forced to fish under unsafe weather conditions because the larger boats were out catching the fleet's quota. To some degree, the fishery was competitive because of the income and status that the most successful or highliner fishers enjoyed. Trip limits did hold back this aspect of competition. However, the fleet quotas created a situation of inherent competition because, if a skipper had poor catches early in the season, he would never be able to compensate later on once other vessels had exhausted the fleet allocation. Hence, a more radical change in management tactics was contemplated during 1983. The plan was to switch from an overall fleet quota to an enterprise allocation, or "boat quota," form of management.

In theoretical terms, enterprise allocations are seen by advocates as a means of avoiding the "tragedy of the commons" often thought to be inherent in common property resources (Gordon 1954; Scott 1955, 1979; Hardin 1968; Larson and Bromley 1990). This tragedy is the overcapitalization and eventual overexploitation that results from individuals competing to harvest as much of a common property resource as quickly as possible to avoid losing the resource to competitors. Allocating a certain amount of the resource to individual enterprises introduces a degree of private property ownership. This is thought to reduce the need to maximize short-term harvests and the overcapitalization that results as each producer strives to increase her or his share of the total. Ideally, this limitation on investment reduces the tendency to overexploit the resource and may even produce greater interest in long-term conservation. Enterprise allocations are also thought to have the potential to reduce administration costs, and fees for allocations can be directed by the state to help pay for these costs. To the degree that enterprise allocations are transferable—that is, subject to being bought, sold, leased, divided and inherited—market forces may also be able to adjust concentrations of labour and capital to fluctuations in the resource (see Chapter 5). Still, as McCay and Acheson (1987) point out, the general theoretical expectations of introducing private ownership into common property resources do little to predict the exact consequences of specific programs in particular settings. Whether or not enterprise allocations will fulfil any or all of their potential goals depends on numerous factors about the resource, the harvesting industry, the exact details of the program, and the degree and means of enforcement of the program's policies (Copes 1986).

The enterprise allocation program for northwest Newfoundland draggers, slated to begin in 1984 and originally proposed for three years, divided the fleet into six vessel classes and allocated each licence holder a yearly quota determined primarily by the length of their vessel. It also split the fleet into four sections, each with its own sector quota that was not transferable to other sectors. The official goals of this particular enterprise allocation program were to:

1. improve the quality of fish landed by the fleet;
2. extend the length of the harvesting season, thus maximizing employment potential in the processing sector;
3. reduce the competitiveness that existed in the fishery in 1983 and reduce the risk of small vessels competing with larger vessels;
4. assist vessels within the fleet to become economically viable over the long term; and
5. reduce or eliminate fish glut which permitted large quantities of fish to be trucked and processed outside western Newfoundland (Canada 1983:1–2).

It is interesting to note that this list of goals does not include several of the potential effects of enterprise allocations. First, although the enterprise allocations were expected to spread harvests over a longer season, lack of transferability greatly reduced the capacity of market conditions to influence the flow of capital and labour in and out of the industry (Boyd and Dewees 1992). Specifically, although concern about the economic viability of the fleet was a major motivation behind the new enterprise allocation system, the system was not designed primarily to allow market forces to reduce the number of vessels in this fleet. The reduction of the fleet, which was seen immediately following the addition of new entrants in 1982 as the "last and most drastic move," was now hoped to be a secondary by-product of the enterprise allocation system. This expectation was based on the belief that the allotted quotas were inadequate for the survival of the 10–20 percent of boats in each category that were encumbered by high debt payments.

From the perspective of 1996, an even more striking aspect of the 1984 enterprise allocation plan was the lack of any reference to the long-term conservation of the stock and the creation of greater conservation consciousness among the skippers (see also Crowley and Palsson 1992:15). In contrast to New

Zealand (Crothers 1988), concern about the long-term viability of the Gulf cod stock played little or no role in the decision to implement the northwest Newfoundland enterprise allocation system. One provincial fishery officer stated that, although the DFO talked about conservation at this time, its actual concern about ecological factors was only "between one and two on a scale of ten." A dragger skipper shared this view and stated that the DFO had "never been concerned about fish, just who catches it." While this is clearly an overstatement, the implementation of the enterprise allocation system was primarily a matter of who would catch the fish.

Although there was considerable debate among the fleet's skippers, the new plan met with substantial support from both the union (see Chapter 4) and the skippers (reported to be around 80 percent in favour). Many fishers, especially those on smaller vessels, liked the plan because it would reduce the need to fish in bad weather. All of the skippers also saw that temporary breakdowns would no longer be economically disastrous. The main reason for support, however, was that the draggers enjoyed a fairly good year in 1983, and the proposed boat quotas, ranging from a low of 350,000 pounds to a high of about 860,000 pounds, were seen as adequate for most vessels. These quotas were also expected to increase over the following few years.

The greatest opposition to the plan came from some of the largest, most successful vessels whose harvesting abilities greatly exceeded their new quotas (see also Dewees 1989). Several fishers and fishery officers later speculated that the eventual support of the new plan by these skippers occurred only because they calculated that they would be able to circumvent the quota restrictions. For example, although enterprise allocations were not made officially transferable until 1990, one skipper recalled that "Actually they were always transferable. We could always find ways to transfer fish to someone else. There was no way they [the DFO] could stop us." Skippers also foresaw an even more important way around the new restrictions: unreported, "under the table" sales. Such sales became common after the implementation of the enterprise allocation program.

The state's role in implementing enterprise allocations at this time could be considered "pro-dragger," primarily because this fleet was seen as the most effective fishing fleet by size in the province. The public outcry from those dependent on the fixed-gear fishery was negated by the fact that they also had a relatively good year in 1983, although a gradual decrease in their landings continued. The state's support for the dragger fleet, however, was ambiguous, especially for those vessels that would struggle under the new enterprise allocation system.

The initial effects of the enterprise allocation system seemed to be primarily positive. The program appeared to stabilize the fishery and reduce glut periods (see LeDrew 1988:18, Fig. 5). This was particularly beneficial to the small independent fish plants that were now able to stay in operation for longer periods of the year. The vast majority of skippers enjoyed the reduced competition and felt it made the fishery much safer. Many also appreciated having a much more certain income on which to base their yearly budgets. Most fishery officials felt that, with slight modifications, the enterprise allocation system would be a continuing success.

Within a few years, however, it became clear that the enterprise allocation system had also brought with it a number of problems that would have profound long-term implications for both the ecology of cod and the social structure of northwest Newfoundland communities. Foremost among these problems was the misreporting of landings, locally known as "under the table" sales. As one skipper stated, "That's when under the table sales started. Because it wasn't necessary [before the enterprise allocation system]. There wasn't any point in it. The real reason was to be able to catch more fish if it wasn't reported." In other words, the boat quotas provided an incentive to underreport sales because a fisher's season would end once the official allowance for the boat had been recorded. Unreported sales effectively extended the quota (Copes 1986; Dewees 1989; Boyd and Dewees 1992). Although it is impossible to know the exact extent of misreporting, skippers will generally concede that all vessels did it "a bit," but most emphasize that a few of the larger boats may have doubled their legal annual catches. Estimates by fishery officers of unreported sales during the mid-1980s usually ranged between 20 percent and 40 percent of the total legal quotas for the entire fleet (personal interviews).

"Under the table" sales were impossible without the cooperation and collusion of buyers because "for a fisherman to cheat, someone has to buy the fish" (Boyd and Dewees 1992:186). The majority of buyers on the coast engaged in the practice for several reasons. First, the enterprise allocation system had greatly reduced the glut problem, with the result that plants could often extend their periods of operation if they could obtain greater supplies of fish. Another reason was that lack of monitoring made it a very easy procedure:

> First when this thing [the enterprise allocations] came into effect, there was no monitoring. No one wasn't watching anything. Not even fishery officers wasn't watching. They just let it go. Suppose you brought in 50,000 pounds of fish. Well you'd just say to the buyer "Why don't you receipt me for 30 [thousand pounds]." . . . It was very easy to do.

There was also the fact that since the practice had become, in the words of one skipper, "the accepted thing," any buyer who refused to engage in the practice would be at a decided disadvantage in attracting vessels. Several skippers reported that the Fishery Products International (FPI) plant in Port au Choix lost considerable fish to the independent plants during this period because of FPI's refusal to engage in unreported sales.

The ability to misreport catches meant that the enterprise allocation system failed to eliminate completely the competitive fishing practices that were so dangerous. Indeed, a "load 'n go goldrush" mentality still often characterized the fishery, especially during the winter when extremely large tows were still occurring. Several accidents probably resulted from such fishing practices and many skippers still marvel that more tragedies did not occur.

The lack of monitoring of sales by the DFO just alluded to may also be relevant to evaluating the role of the state in supporting different sectors of the fishery during this period. If one ignores the accompanying misreporting, the enterprise allocation system appeared to be a state policy that "made it more difficult for operators of large vessels with correspondingly high mortgages to keep afloat, but has eased the burden on smaller enterprises" (Sinclair 1986:116–17). However, many skippers feel the state passively condoned "under the table" sales. When asked by one of the authors (Palmer) if the government fishery officers knew about these sales, one skipper replied, "Oh my God, Craig boy, they had to know. They had to know that fish was sold under the table. I mean if you [only] had 100,000 pounds of fish [an amount that could be easily caught in a few weeks] to catch and fished all summer, God, you had to know." This passive acceptance of "under the table" sales might be seen as unofficial support for the larger vessels, who engaged in the practice to the greatest extent. At least this appears to modify the view that the enterprise allocation system could be seen as a state policy against the larger draggers.

In addition to the misreporting of sales, the enterprise allocation system was also linked to high-grading fish and discarding smaller, less valuable fish (Boyd and Dewees 1992). Since a skipper was limited to a certain weight of fish, it was crucial to maximize the value of the fish sold. Although it is again impossible to know the exact extent of such discarding, it reportedly occurred both at sea and at the wharves of the fish plants. As one skipper stated, after praising the enterprise allocations for reducing the intense competition, "the only bad thing was that it created a lot of cheating."

Although fishing was fairly successful during 1984, limited primarily by poor weather conditions, this year also saw the beginning of several other trends and practices not directly related to the new enterprise allocation system that would come to dominate the future of the fishery. Foremost among these were the reports of cod becoming smaller and slightly more difficult to find. Although many gill-net fishers had complained since the late 1970s about the necessity of using ever smaller sizes of mesh to catch the diminishing size of fish, most dragger skippers pointed to the mid-1980s as the time they first noticed the fish getting smaller and scarcer.

The response of the dragger fleet to this evidence of decline in the stock was to place "liners" in their nets. Using a liner involves placing another layer of net into the cod end (the section of the net where fish are trapped) in order to reduce the mesh size and retain more of the smaller fish. One dragger fisher explained "People took what they call the 'belly'—the same size mesh—and just put it in their net and leave it in and let it fall back through so it criss-crossed the cod end." Although a few skippers may have used liners previously, the mid-1980s saw a dramatic increase in this practice.

This widespread use of liners provides insight into the perception of the state of the cod stock on the part of fishers and fishery officials during the mid-1980s. When asked if there was a lot of talk about the health of the stock and the possible destructive consequences of misreporting and using liners, one skipper stated: "No, that wasn't really common. That wasn't the normal thing at that time." It would, however, be inaccurate to state that illegal fishing practices were adopted because dragger skippers were unaware or unconcerned about potential ecological consequences. Indeed, understanding the reasons for the adoption of these practices requires appreciation of the social environment within which the fishers of this region operate.

Pervasive in explanations for the use of liners are references to the greater catches of other skippers. Indeed, the competitive nature of this fishery, even after the enterprise allocation system was implemented, was given by one skipper as the *sole* reason for the origin of liners: "I'll tell you when liners started; it was simple. It started when one fella came in with fish and the other didn't." Social pressure to match the catches of other skippers is also almost invariably found in the reasons skippers gave for adopting the practice once it was being used by others in the fleet. One skipper recalled how he first became aware of the use of liners:

> We could have caught our fish a lot easier with a liner, but the first two years [1984 and 1985] we didn't even know they were using it. We just figured, boy, there was something wrong with us, that we were doing something wrong. Then one day we were told, we didn't figure it out until we were told. . . . We just thought we were stupid.

A similar story was given by another skipper:

> There was people who just went clean out of their minds, spend days, weeks, at their nets. . . . Guys coming in with fish and they can't get none—take off the net, change this, change doors, you know? Then all of a sudden somebody would finally tell them what was going on.

Nearly invariably, once the cause of the discrepancy in catches was discovered, liners would be adopted: "Now my name won't be on this, right? . . . Well, then, I cheated [by using liners and misreporting] like the rest of them, but I was forced into it by watching others make more money [by cheating]." Within several years this had made the use of liners, in the words of one skipper, "the standard thing."

In addition to not wanting to be outdone by one's competitors, the fear of detrimental social consequences often prevented the voicing of concerns about conservation of the stock. One skipper stated, "We should worry about what's gonna be left for our sons, but when I told one fella [dragger skipper] this, he just said, 'Fuck them all.'" The ability of such negative social responses to prevent the voicing of ecological concerns was also described by another skipper who stated, "A lot of the fishermen were concerned over the small fish, but the government, the fisheries, didn't seem to take any advice from anybody. All you did by saying anything was make bad friends."

The last quote refers to the lack of concern and enforcement by the DFO, a point that also nearly always occurred in discussions over the causes of illegal fishing practices. The following conversation typified the view of dragger skippers regarding the DFO's attitude towards liners in the mid-1980s:

Dragger skipper: I had an old boat and it wasn't much different whether I used a liner or not I wasn't catching much fish with it. But nobody seemed to care! The fisheries didn't care about it.

Interviewer: They knew about it?

Dragger skipper: Oh yes, they knew about it, but nobody said anything about it.

Interviewer: How did they know?

Dragger skipper: The liners were left in. Just walk down to the wharf.

Interviewer: They'd be easy to see?

Dragger skipper: Oh, definitely easy to see. The fisheries would just turn a blind eye. . . . It wasn't until the late [19]80s that liners was done more discreetly, you know [then] it was covered up.

Explanations for the adoption of illegal fishing practices typically state that the individual did not want to adopt the practice but could not stand being outfished by competitors who did, and hence they blame the DFO for not forcing these competitors to follow the rules. Although such explanations obviously use the DFO as a scapegoat to avoid placing blame on one's self and one's peers, the attitude of the state towards the use of liners may provide insight into the state's relative degrees of concern about the economic viability of the fleet versus the conservation of the stock at this time.

Despite realization that the enterprise allocation system had inadvertently led to illegal and destructive fishing practices, the program, with modifications to allow a slight increase in the temporary leasing of licences, was extended for another three years (1987–89). It was felt that the illegal practices that had accompanied it would be dealt with by increased DFO monitoring of the fishery. The difficulty of this solution was, however, made readily apparent during the 1987 winter fishery when a number of dragger skippers circumvented the new monitors simply by starting to fish ten days earlier than scheduled. Such practices also nullified a new regulation limiting draggers to only 50 percent of their year's catch in the winter fishery, a regulation implemented to relieve the increased fishing pressure on this area which had resulted from the growing scarcity of fish in the rest of the Gulf.

Other Dragger Fisheries

Mobile-gear vessels also fished for other species when their skippers possessed the necessary licences. Shrimp was particularly important, but shrimp landings declined in 1984 and 1985. Although this was partially due to bad weather, there was also a reduction in the effort directed at shrimp because, compared to an equally valuable catch of cod, a shrimp catch required more time and expense. At the time, this decline in effort was not cause for concern because there was some talk about the shrimp stocks being overfished. There was also growing anxiety about the destruction wrought by shrimp dragging on other species such as redfish, flounder and cod. This was because the small mesh size needed to catch shrimp could also catch large quantities of juvenile members of these species. Reports of "the ocean turned red" during the shrimp fishery by the huge amounts of dead baby redfish discarded from shrimp nets had circulated since the 1970s, but ecological concerns still remained largely in the shadows of economic issues. For example, in 1982 the large numbers of dead redfish that Sinclair observed during shrimp fishing trips seemed of no concern to the fishers, who appeared to take this sight for granted.

As the cod quotas began to decline in the mid-1980s, and the DFO attempted to place more restrictions on the dragging of cod, shrimp once again increased in importance. Shrimp catches were up slightly by 1986 as a direct result of the decline in cod quotas, not because this fishery had become more profitable. The price of cod increased in 1987, but the price of shrimp increased more dramatically, jumping from 60 to 90 cents per pound. This price increase, combined with lower cod quotas, increased tension between shrimpers and non-shrimpers.

A new prey species also began to play a small but growing role in the dragger fishery during this period. In 1984, the DFO began to encourage skippers to harvest redfish to supplement their yearly incomes. Considerable quantities of redfish were suspected to gather off the southwest coast in the winter and to disperse throughout the Gulf during the summer. But two factors kept interest in this fishery low during the next few years. The first was the low price paid for redfish. The second was the fact that during the winter, when redfish were concentrated in large schools, they were found higher in the water column than the otter trawl nets could easily access. However, the problem of low prices diminished, especially for boats lacking shrimp licences, as the lucrative alternative of cod dragging was further restricted by falling quotas and better enforcement of regulations (see Chapter 5). In the mid- to late 1980s, a few of the more enterprising dragger skippers solved the second problem by equipping new boats with expensive "mid-water" trawling capabilities. By 1988, redfish had become an integral part of the fishery, especially during the winter, with the mid-water trawl boats having the greatest success.

The scallop fishery is exploited occasionally by draggers and on a more regular basis by a small number of longliners who lack otter trawl licences and combine scallop dragging with gill netting for cod. The scallop fishery boomed after 1984 as the price jumped from $5 to $7 per pound. Although scallop licences had been frozen in 1985, this was not a problem during the next few years. Even when the scallop fishery was in relatively good shape in 1986, only sixty to seventy of the licences were being fished. Following a dramatic price drop that started in 1987, down to $5 per pound or less, there were only thirty to forty active scallop fishers by 1988. Hence the scallop fishery declined with the price drop at precisely the time it was most needed as an alternative by both dragger and gill-net cod fishers.

Fixed-gear Fisheries

Domestic commodity producers who were still dependent on the fixed-gear fishery saw some declines in catches between 1982 and 1987 (see Figure 1), but it was not until the end of this period that their plight began to seriously affect management decisions. One of the reasons that the declining catches did not cause more social protest is that most fixed-gear fishers still managed to obtain the minimum ten weeks of employment necessary to qualify for Unemployment Insurance Compensation (UIC). As catches continued to decline, the attainment of enough "stamps" (insurable weeks of employment) to qualify for UIC began to take on primary significance (Hanrahan 1988; Kennedy 1996:108–09; Palmer 1992). This was reflected in the phrase "fishing for stamps." The presence of UIC is one reason that, despite the decline in catches, most attention directed towards problems in the fixed-gear fishery during the early part of this period was focused on the issue of gear conflict, not declining catches. Even these complaints were seen as minor problems that did not necessitate new regulations that might unduly hinder the operation of the dragger fleet. Indeed, the first major new regulation implemented in 1983 was to limit entry into the fixed-gear groundfish fishery.

By the mid-1980s, the decline in the fixed-gear fishery could not be ignored. Despite temporary periods of good fishing, and the high prices of 1987, gill-net fishers, who had been steadily reducing the size of their mesh since the 1970s, were clearly encountering more difficulty in finding fish, with the fixed-gear fishers of the southwest coast suffering the greatest reductions in catches. Along the northwest coast, the cod-trap fishery, which had been the backbone of the fixed-gear fishery for decades, ceased to be a reliable source of income by 1984 and was simply "dead" after 1987 (but see Chapter 5).

Throughout these years, lobster fishing benefitted from reasonably high prices. Hence, lobster continued to be a partial saviour for fixed-gear fishers with lobster licences, but limited entry had made the lobster fishery unable to take in significant numbers of new participants. The herring fishery, although highly variable, was in a state of overall decline, while the salmon fishery was in such serious decline that its eventual end was beginning to be sensed (see Figure 1).

The bright spot in the fixed-gear fishery was capelin. Spurred by new demands from the Japanese market, this fishery boomed in the mid-1980s. This was particularly important for fixed-gear fishers without lobster licences. There were, however, problems with this fishery. First, the season was extremely short, lasting less than three weeks under the best conditions. Second, it would prove to be very unpredictable from year to year as it hinged not only on the presence of capelin but on the size requirements of Japanese buyers. Finally, it was an inherently wasteful fishery as the Japanese buyers were only interested in the roe from female capelin, leaving the male capelin to be dumped.

The difficulties in the fixed-gear fishery finally led to the formation of the Newfoundland and Labrador Fixed Gear Association in 1987. This group did not call for the abolishment of the dragger fleet, but only for regulations that would keep it from destroying the fixed-gear fishery. It became quite vocal in 1988, focusing primarily on public relations and education while attempting to influence the DFO to place further restrictions on the draggers. The draggers responded to the fixed-gear fishers by forming their own Otter Trawler's Association to offset what they felt was unjustified negative publicity and to protect their own interests in the DFO's decision-making procedures. The social conflict between fixed-gear and mobile-gear fishers that had existed informally since the origin of the draggers had become institutionalized.

Fish Plants

To understand the opinions of local residents and state officials towards the dragger fleet, it is necessary to understand the relationship between the draggers and the local fish plants. Once the traditional "making" of fish had been replaced with fresh-fish processing, fish plants became a major social and economic factor in the lives of northwest Newfoundland residents. By the early 1980s, fish plants had become the largest employer in the area. In 1984, fish plant work provided 4,000 person-years of employment (Canada 1990:14). Fish plants supplied steady work during the summer, with some plants also processing fish from the winter fishery, and fish plant work became a reliable means of qualifying for UIC.

This booming processing industry became increasingly dependent on the fish caught by the dragger fleet. Indeed, it was the overabundance of dragger fish during the early and mid-1980s that made the glut during the peak summer fishing months into the main problem facing northwest Newfoundland fish plants. Millions of pounds of fish landed on the Great Northern Peninsula had to be trucked to other areas of the island for processing because local plants had already exceeded their capacities. This led to the enterprise allocation system to reduce the glut problem, which led to the widespread adoption of "under the table" sales, which in turn contributed to decreases in the size and quantity of cod. This encouraged the use of liners, reductions in quotas and further social conflict. To understand the role of the processing sector in these changes, it is necessary to know something about the social context of fish plants along the northwest coast of Newfoundland.

As the local fish plants became dependent on the dragger fleet, one of the most frequently heard rebuttals by dragger fishers to calls for their abolishment was that "if the draggers go, the fish plants go, and if the fish plants go, then everybody goes." Although fixed-gear fishers could reply that this situation was caused by the draggers destroying the fixed-gear fishery in the first place, the draggers' position coincided with the current reality.

Not only did the processing sector's dependence on the dragger fleet influence the state's support for the draggers, it also complicated the social conflict between mobile-gear and fixed-gear fishers. Perhaps the greatest complication came from the fact that local community fish plants employed many women, most of whom had not earned wages previously. In addition to changing the social interaction patterns of many women, fish plant employment had a dramatic impact on the economic structures of households. One woman recalled the impact of women having their own money: "Yes, it was nice, you know, because a woman would want to spend money on different things than a man would." This change in the internal dynamics of household economics and politics created cross-cutting ties within communities and families, because many of the wives of fixed-gear fishers worked in fish plants processing fish caught by draggers. Hence, as their own catches began to decline, fixed-gear fishers often found their households growing increasingly economically dependent on the fish caught by the draggers. Whether this tempered or aggravated feelings of resentment towards the dragger fleet is open to debate.

What was clear was an expectation that fishers, particularly draggers, should be "loyal" to the communities where they lived (Palmer 1992). Being "loyal" means selling fish to the owner of the home community's fish plant. Although most dragger skippers agreed that supplying one's home community with fish was the "proper thing," there were other factors to consider in the decision of where to sell fish. In addition to differences in price, it was also important to develop a relationship with certain buyers to ensure a place to sell one's catch during certain periods of the fishery. For example, because only two companies bought shrimp, many skippers sold their early-season cod to these companies in order to ensure a buyer for their shrimp later in the season.

The fact that some companies owned fish plants in several communities meant that it was also sometimes difficult for skippers to be loyal to their home communities even when they wanted to. Fish sold to one plant might end up being processed at another because the company might decide to truck the fish to a different community.

How fish-processing companies' decisions interacted with the decisions of fishers in determining where fish was to be processed is illustrated by the following example. Despite a stated policy of dividing fish equally between two fish plants owned by the same company in two neighbouring communities, the residents of one community were convinced that they did not receive their "fair share" of fish. This sense of unfairness was exacerbated by the fact that the company also processed its shrimp in the same community that was felt to already receive more than its fair share of cod. The main explanation given by local residents for this imbalance was that over a dozen dragger skippers, most of whom had shrimp licences, lived in the "favoured" community. In contrast, only one dragger skipper, who lacked a shrimp licence, lived in the other community. Residents talked about this one boat being unable to "supply their plant" and, in addition, the lone skipper was unable to influence the company by threatening to sell his fish to another company if fish were not delivered to his own community. The dragger skippers in the other community, however, were able to make this threat on a fairly regular basis to insure that fish were delivered to their own community and to maintain prices at acceptable levels.

The role of women in the fishery was related to the explanation often given by residents for why dragger skippers were not as loyal as they might be. During the early 1980s, many wives of dragger fishers worked in their home community's fish plant. Hence, draggers who were "loyal" to their home communities were also contributing to their own household income. By the late 1980s, dragger skippers started to form "companies." The main advantage was that skippers and crew members could draw unemployment payments during nonfishing weeks throughout the year because they were now technically "employees" of the company instead of co-adventuring fishers. Another advantage was that skippers and crew members could list their wives as employees of the company and they could also receive labourer UIC payments. The fact that the wives of dragger fishers became employed by their husband's companies instead of the local fish plant was sometimes seen by local residents as causing the husbands to be less loyal to the community. Although many wives did make significant contributions to their husband's companies by keeping books, baking bread, running errands and helping with repairs, further resentment towards the draggers was engendered by the view that many of these wives received UIC without really contributing to the enterprise. As one fixed-gear fisher stated: "Hell, lots of those women don't even know the colour of their husband's boat."

Conclusion

The period between 1982 and 1987 may be seen as one of continuing expansion of the dragger fleet and ambiguous state policy, as the DFO struggled with growing evidence of overcapacity. It was also was during these years that perhaps irreparable damage was done to the cod stock. It is tempting to view this damage primarily as the result of the failure of the enterprise allocation system, which was the main regulatory device applied to the dragger fleet during this period. Although it is true that the enterprise allocation system reflected a focus on economic concerns and too little appreciation of the ecological precariousness of the cod stocks, the problem appears located more in the failure to enforce regulations than in the regulations themselves. In addition to overly optimistic perceptions of the health of the cod stock, the misreporting, discarding, and use of liners were prompted by the social environment within which the fishers operated. Responsibility for the destructive practices must obviously be shared in some measure by the dragger fishers, plant owners and the DFO. In any case, by the end of this period, what had been an economic issue concerning the viability of the dragger fleet had been transformed into an ecological-economic problem that divided residents and threatened the survival of everyone on the peninsula.

Note

1. Skippers still talk in pounds and feet. One never hears mention of a 10,000-kilo tow (1 kilogram = 2.2 pounds). Hence we have retained the common usage rather than translating into metric measures.

4
Social Divisions and Social Cohesion

By 1988 the northwest Newfoundland fishery was clearly in decline. Fixed-gear catches were plummeting and the mobile-gear fleet was experiencing a period of diminishing quotas. The issue of the economic viability of the draggers was now embedded in the much larger issue of the survival of both the Gulf cod stock and the residents of northwest Newfoundland. The next five years would be ones of formal and informal conflict between the different sectors of the fishery as they struggled to influence state policies in directions that would allow their survival. Indeed, the question of whether or not the future fishery would include an inshore fixed-gear sector had now become the key concern of local residents. Many residents felt there would not be enough fish for both a fixed-gear and even a reduced mobile fleet. They were also aware that the decision about which fishery would survive would determine which current residents would be able to continue to live in the area. Hence, this period, which would form a crucial test of the relations between the state, domestic commodity producers and the small capitalist dragger fleet, was also a time of social divisions. Instead of merely considering these social divisions as results of state policy, it is crucial to appreciate their potential role in influencing state policy. Indeed, many of the events of the late 1980s and early 1990s can only be understood within the context of the social relations among northwest Newfoundland residents. Hence, an overview of the social context of the area during this period is presented before the specific fishery events that took place between 1988 and 1992 are examined in Chapter 5.

In describing the egalitarian ethic that pervaded northwest Newfoundland communities in the early 1960s, Firestone stated that this ethic was largely the result of everyone being in "the same boat" (1967:viii). The rise of the dragger fleet and the differentiation of the fishery meant that by the mid-1980s this fundamental basis for egalitarian social relations no longer existed. The role of such developing social divisions in determining the fate of domestic commodity production has often been overlooked. In arguing that such neglect is unwarranted in the case of the northwest Newfoundland fishery, Sinclair wrote:

> Furthermore, the concentration of harvesting capacity in a small dragger fleet, protected by public policy, has generated resentment and opposition from other fishermen and has given rise to open conflicts, *to which theories about the fate of petty primary producers accord little attention* (1985:146, emphasis added; also see McCay and Creed 1990).

Sinclair also pointed out that the need to examine the role of social conflict in determining the future fate of the dragger fleet and DCP in northwest Newfoundland would probably increase because "economic concentration and community polarization might become even more extreme than they are currently . . . [and] [c]onflicts . . . among classes of primary producers . . . are likely to appear as important components in the process of change" (1985:146).

This prediction of increased social conflict was based primarily on the expectation of continuing economic disparities between different sectors of the fishery. Although this expectation proved to be correct, an even greater factor began to feed the fires of social conflict during the late 1980s. By 1990, not only were northwest Newfoundland residents in very different "boats" both literally and figuratively, but many of those dependent on the small fixed-gear boats were also blaming those dependent on the larger boats of the dragger fishery for destroying the cod stocks. Yet, despite these divisive forces, much of the social fabric of the northwest Newfoundland communities remained intact. Recognition of these strained but enduring social ties is necessary for understanding the fate of DCP and the dragger fleet during the late 1980s and early 1990s. To comprehend the social ties among the different sectors of the fishery, it is helpful to begin by looking at an institution designed to be, but rarely operating as, a cohesive force in northwest Newfoundland: the Fishermen, Food and Allied Workers Union (FFAW).

The Union

Perhaps the most obvious structural factor that might have counteracted the divisions within the fishery during this period was the FFAW. Despite the existence of a smaller competing union (the Newfoundland section of the United Food and Commercial Workers) that had organized several fish plants, the FFAW brought together fixed-gear fishers, dragger fishers and fish plant employees, ostensibly uniting them in a common effort. In reality the union acted as such a cohesive force only rarely during the late 1980s and early 1990s. More often, the FFAW's ability to lobby for higher fish prices and other policies and generally to influence state policy was controlled by the sector of the fishery that dominated the union. Hence, what could have been a source of cohesion was often a source of conflict.

The unionization of fishers in the area dates back to the merger of the Canadian Food and Allied Workers' Union, which had been organizing plant workers and trawlermen between 1967 and 1970, and the Northern Fishermen's Union, which began life in Port au Choix late in 1969 (Sinclair 1985:127; Inglis 1985). At that time it was the young fishers who had introduced longliners to the area who led the way in organization, and throughout the next two decades the more entrepreneurial longliner, and then dragger, skippers played the most active role in the union.

The union was of considerable aid to the rise of the dragger fleet and in this sense acted largely in concert with state interests in modernizing the fishery. Prior to 1981, however, this relationship was more ambiguous because the union played a major role in several organized work stoppages (especially the strikes of 1974 and 1980) that were partially successful in forcing Fishery Products to raise prices. Such conflict and disruption of normal business were not on the agenda of either the provincial or federal governments. In contrast, the union's promotion of licencing, professionalization and new catching technologies did fit well with the federal government's objectives for the industry. Thus, in 1982 the union helped lobby for the expansion of the dragger fleet, and it supported the adoption of enterprise allocations in 1984. Although the union was at least partially in opposition to the state throughout this period, especially in disputes with state-sponsored fish-processing companies, the state and the union both shared fundamental support for the dragger fleet. This congruence of interests was the result of the state's general support for the modern dragger fleet (as opposed to traditional, unproductive open boats) during this period and the dragger sector's dominance of the union.

> In disputes with the fish companies, draggermen have been at the centre of the action. . . . They have become particularly aggressive because they have made large personal investments in their boats, they recognize that the precarious nature of the fishery threatens their future, and they have developed confidence in their ability to act collectively in defence of their interests (Sinclair 1985:140).

As the stocks collapsed in the late 1980s and fixed-gear fishers increasingly came to blame the dragger fleet, the dragger fishers' domination of the union took on added importance. Perceiving the union as no longer serving their interests, many fixed-gear fishers ceased to participate and some formed their own associations (see Chapter 3). Hence, by the late 1980s the union was perceived largely as a powerful lobby group for the draggers. For example, when asked about the effectiveness of the various attempts at forming alternative fixed-gear associations, one provincial fishery officer replied that "all the small-boat [fixed-gear] fishermen could take their boats to St. John's and burn them on the government's doorstep, and they wouldn't get as much attention [from the government] as . . . [an influential dragger skipper] gets by clearing his throat at a union meeting."

This perception of the union as an effective lobbying group for the draggers was becoming less clear by 1990. Many dragger fishers were far from satisfied with the union; they saw it as largely ineffectual and occasionally acting against their interests. For example, it was common to hear the statement that the union's negotiation of a minimum price of fish with the Fisheries Association of Newfoundland and Labrador actually kept prices down. Skippers of smaller draggers were particularly unhappy with the union because they felt it favoured the larger (65-foot) vessels. Even some fixed-gear fishers agreed that the role of the union was changing: "It used to be a dragger's union, now I don't know who it's for."

Although many dragger fishers voiced disapproval of union actions, they also realized that the union retained some power. This was reflected by the greater participation of dragger fishers at union meetings. At one meeting, ostensibly for both types of fishers, a dragger skipper was asked why there were so few fixed-gear fishers in attendance. He replied that they never bothered to get involved, and that many still did not realize that meetings and board rooms, not the open sea, were where fish were actually caught. He then added, "That's why they are still small boat fishermen!"

It is readily apparent from this brief description that, despite its attempts to create common interests among its diverse membership, the FFAW was unable to offset the divisive forces that separated the various sectors of the fishery. These forces were, however, partially controlled by other, more informal, methods. An understanding of these methods requires a more detailed analysis of the social divisions in northwest Newfoundland.

Social Divisions

During 1990, Palmer conducted two surveys (see Appendix) that shed light on the social divisions among northwest Newfoundland residents. Upon arriving in the area in May 1990, he undertook an informal unstructured survey that recorded complaints about the fishery from 112 of the first 230 northwest Newfoundland residents he encountered. These complaints were in response to Palmer's relatively standardized introduction which included the statement that he was interested in peoples' opinions about the causes of problems in the fishery. These responses were then used to help construct a formal structured survey of 171 adults in one northwest Newfoundland community (Palmer's base research site) during the fall of 1990. Together, these surveys reflect both the social strains that were being felt by residents in the area and the tactics used to maintain social ties. Table 5 shows the gender, occupation and responses of the people Palmer met during his initial survey.

The 112 people who made statements about the fishery listed a total of 378 problems; most people listed only one or two problems, but a few referred to more than ten. Table 6 lists the ten most commonly mentioned problems and shows that three types of complaints dominated the fishery. Two of these—complaints about the draggers and the government—were expected. The complaints about the illegal behaviour of local residents were more of a surprise and suggested certain insights into both local social dynamics and the interrelations between the other two types of complaints. Hence, these three most commonly mentioned problems were analyzed in terms of their interconnections and their role in both the social divisions and the social cohesiveness in the area.

Table 5. Statement of Fisheries Problems by Gender and Occupation (Initial Survey)

Gender	*Occupation*	*No. Stating Problems*	*No. Not Stating Problems*	*Total*
Male	Small-boat fisher	50	12	62
	Dragger fisher	17	7	24
	Fish plant worker	5	6	11
	Nonfisher	21	35	56
	Retired	6	6	12
	Total	99	66	165
Female	Small-boat fisher	1	1	2
	Dragger fisher	0	0	0
	Fish plant worker	3	10	13
	Nonfisher	8	35	43
	Retired	1	6	7
	Total	13	52	65
Total		112	118	230

Table 6. The Ten Most Frequently Mentioned Problems with the Fishery

Problem	*No. Stating Problem*
Government (no specification of exact problem)	67
Illegal behaviour of local residents	45
Local dragger fleet (for reasons other than illegal behaviour)	44
Offshore Canadian trawlers	24
Low fish prices	19
The union (FFAW)	15
Raising of the UIC requirement to fourteen weeks*	14
Foreign trawlers	9
Lack of government enforcement of fisheries regulations	8
Seals	6

*The government had recently announced that fishers would have to obtain fourteen weeks, instead of the normal ten weeks, of insurable income during 1990 to qualify for UIC. This was seen as a major indication that the state had decided to get rid of those people dependent on the fixed-gear fishery. Although the requirement was returned to ten weeks at the end of the 1990 fishing season, it made many residents realize that the UIC system could not be taken for granted.

As expected, many (44) residents made specific reference to the local dragger fleet as a cause of problems in the fishery. This evidence of social division between dragger fishers and other residents was confirmed by Palmer's formal structured survey during the fall of 1990, in which 89 percent of the 171 respondents stated that the Gulf cod stock was "almost gone," and 74 percent stated that the local dragger fleet was a primary cause of this decline (Palmer 1992; also Canning and Pitt Associates 1992). Sixty-eight percent favoured closing the winter fishery, and 66 percent supported abolishing the dragger fleet on condition that dragger fishers would be given compensation. All findings were consistent with the fact that conversations around the kitchen tables of families dependent on the fixed-gear fisheries often centred on the greed and illegal practices of dragger fishers. Dragger fishers were also quite aware of this aspect of public opinion and would often begin their own explanation of the problems in the fishery by acknowledging that they knew what other people were saying:

> Well, I can tell you what most people are telling you—"the draggers"—but it's actually just the nature of things, there's always been bad years. It's water temperatures that keep fish from coming in and it's seals and whales that have reduced the stock. We need to educate people that it's not just the draggers.

Other evidence, however, shows that the social context of northwest Newfoundland is much more complex than a simple division between mobile-gear fishers and fixed-gear fishers.

Social Cohesion

The indications of social division between fixed-gear fishers and dragger fishers just presented are misleading in that they suggest a pattern of social interaction of either complete segregation or open conflict. Although there was variation among communities, in general, such patterns were not found. For example, when the 171 surveyed residents were asked how the problems in the fishery over the last three years had influenced how people in the area get along with one another, only 27 percent said that they had been "split apart," while 66 percent stated that the problems in the fishery had had "no influence" (3.5 percent said they had been "brought together," and 3.5 percent said they did not know). This is consistent with the fact that, although there would be a "scattered fight" between intoxicated fixed-gear and mobile-gear fishers at local clubs, such incidents were uncommon. The extent to which social relationships remained cooperative despite the underlying tension was often noticeable to local residents. One Port Saunders resident, asked if his community was divided between fixed-gear and dragger fishers, responded by saying:

> No sir, you might think so, but it's not. I mean . . . [a fisherman] who's the biggest kind of critic of the draggers, might be sitting here telling you how those damn bastards [dragger skippers] are destroying the fishery, and then . . . a dragger skipper with his car stuck in a snowbank comes in, and it's "why sure, boy," and out the door to give him a push!

This continued cooperation is probably the result of several factors. First is the long history of past cooperative endeavours produced by the myriad strands of kinship and common residence. Second, many fixed-gear fishers will, in certain social settings, state that they probably would have done the same thing that the dragger fishers did if they had had the chance. Indeed, there is even an occasional hint of respect among some fixed-gear fishers for their neighbours who took the "risk" of getting into the draggers back in the 1970s: "You know, I'm the same age as . . . [a successful dragger skipper] and we both thought about getting into a dragger back when they started. But I just couldn't see going into that kind of debt. We just weren't use to that kind of risk." Perhaps the most important reason for the continued social cohesion of the area is simply that residents are currently so entwined in social relationships that crosscut the divisions of the fishery that life would be nearly impossible if everyone in a different sector of the fishery were to be shunned. Further analysis of the other two most common responses to the initial survey suggests some of the methods by which such relationships are preserved in spite of divisive forces in the fishery.

Although common, complaints against the local dragger fleet were actually less common than those against the government (sixty-seven people) and about the illegal activities of local residents (forty-five people). Hence, these responses were further analyzed to identify the interconnections between these complaints and their significance for social divisions in the area. Sinclair (1983) and Copes (1986) point out the importance of local attitudes towards illegal fishing activities in the implementation of fishery regulations, and McCay and Creed (1990:221) found that fishers often take "the cheating issue seriously." Hence, the complaints about illegal activities were analyzed to see if these responses were simply another manifestation of antidragger sentiment on behalf of nondraggers, or if they suggested more complex social divisions. Table 7 presents the number of people complaining about illegal activity by their occupation and whether or not the complaint specified dragger fishers as the perpetrators of the illegal behaviour.

Table 7 indicates that only a minority of complaints about illegal behaviour were directed exclusively at the local dragger fleet. The other complaints included illegal practices by fixed-gear fishers such as selling of undersized fish, cheating on unemployment insurance benefits (UI) and misreporting of sales by fixed-gear fishers and fish plants. The government had recently announced that fishers would have to obtain fourteen weeks, instead of the normal ten weeks, of insurable income during 1990 to qualify for UI. This was seen as a major

Table 7. Perceptions of Sources of Illegal Activity, Draggers and Others

Occupation	*No. Stating Problem*	*No. Stating Only Dragger Personnel Behaved Illegally*	*No. Stating Only Non-dragger Personnel Behaved Illegally*	*No. Stating Both Categories Behaved Illegally*
Small-boat fishers	51	4	10	2
Dragger fishers	17	0	2	4
Nonfishers	29	3	4	9
Fish plant workers	8	1	4	1
Retired	7	0	1	0
Total	112	8	21	16

indication that the state had decided to get rid of those people dependent on the fixed-gear fishery, and it remained a major topic of conversation throughout the summer of 1990. Although the requirement was returned to ten weeks at the end of the 1990 fishing season, it made many residents realize that the UI system could not be taken for granted (see Chapter 6).

Hence, although the large number citing illegal behaviour by local people as a problem in the fishery suggests the presence of social divisions, these divisions are not simply between dragger fishers and other residents. Indeed, complaints about illegal behaviour, especially concerning UI fraud, were also frequently levelled towards people not involved in the fishery. Instead of a simple division between fixed-gear and dragger fishers, the crisis in the fishery appeared to have contributed to a more diffuse type of conflict among residents.

The pervasive accusations of illegal activities on behalf of one's neighbours are reminiscent of Sider's (1986) argument that an exploitative merchant

Table 8. Perceptions of Responsibility for Problems

Occupation	*No. Stating Problem*	*No. Blaming Only Illegal Behaviour*	*No. Blaming Only Government*	*No. Blaming Both*
Small-boat fishers	51	5	14	11
Dragger fishers	17	1	7	5
Nonfishers	29	6	10	10
Fish plant workers	8	3	3	3
Retired	7	1	4	0
Total	112	16	38	29

capitalist system had led Newfoundland fisher families into mistakenly thinking they themselves—instead of the exploitative governmental system under which they lived—were the authors of their own problems. Table 8, however, suggests that this does not seem to be the case in the current situation because conflicts among residents manifested in complaints about illegal activities were often combined with charges against the government. Indeed, it shows that people of all occupations often made a connection between the illegal activities of local people and the failings of government.

Analysis of specific responses suggests that complaints about the government often served to temper or even displace animosity towards one's neighbours. Usually, however, only part of the blame was diverted to the government. For example, when one dragger skipper was asked if he felt small-boat fishers blamed him for the decline in the fishery, he replied, "No, because it's not the people's fault, it's the government's fault, and it's too bad because it does cause hard feelings between people." Another example of a partial deflection of blame is the statement by a fixed-gear fisher that "the problem is the damn draggers destroying the small fish with those liners. The government should really stop them from doing that."

Complaints about offshore draggers and seals, regardless of their validity, also served a function of lessening social conflict between dragger fishers and their neighbours. A poignant example occurred during a Christmas celebration when a group of fixed-gear fishers and dragger fishers started to discuss the conflicts between them. The potentially tense situation was defused when they joined in the singing of the song "Seven Spanish Trawlers," which blames the problems of all Newfoundland fishers on foreign offshore draggers.

In addition to diverting blame to other causes when talking to members of other sectors of the fishery, social relationships are also maintained by the tactic of simply not talking about the fixed-gear/dragger conflict with members of the other sector (see Canning and Pitt Associates 1992). This tactic often goes by unnoticed in everyday interactions, but it became poignantly clear at one 1990 meeting of all types of fishermen held by the DFO to discuss the future of the fishery. A lively and cooperative discussion prevailed despite the DFO representatives presenting disturbing figures such as the fact that the percentage of fish caught in the region by mobile-gear had grown from 14 percent in 1975 to 73 percent in 1990, while the fixed-gear percentage had declined from 83 percent to 24 percent during the same period. Even the result from a cost and earnings survey showing that the average mobile-gear annual income had dropped from $109,000 in 1987 to $32,970 in 1989, while the average fixed-gear annual income had dropped from $8,000 to $2,036 during the same period, failed to create any acrimony. When the issue of the conflict between mobile-gear and fixed-gear fishers was explicitly presented for discussion, however, the room became absolutely quiet (Palmer 1992:26). After several tense moments, the normal social atmosphere returned when someone shifted the topic to overfishing by "foreigners." One fisher stated after the meeting that the DFO officer in charge had lived in the area long enough to know better than to bring the subject of mobile-gear/fixed-gear conflict up for discussion because "once we start talking about that, we'll be at the knuckles for sure" (Palmer 1992:26).

Although these various tactics have allowed social life to continue despite the divisive forces of the fishery, residents are still quite aware of the underlying strain and its origin in the mobile-gear/fixed-gear conflict. One ex-dragger skipper ended his discussion of the future of the dragger fleet (Palmer and Sinclair, forthcoming) by saying:

> It can't go on like it's been going on. Because you can't have two types of fisheries [small-boat fixed-gear and draggers]. There's just too much conflict for one thing. That's what I find about it; you can't have a real solid community kind of place to live in with the two types of fishery we got. I don't think it can be.

Conclusion

It appears that both the social divisions and the muting of these divisions have influenced the state's fishery policy. On the one hand, the failure to discuss controversial topics has prevented residents from reaching any compromises that might allow them to lobby state policy with a unified regional voice. On the other hand, the need to maintain social ties despite the divisive forces of the fishery has probably reduced the effect that social protest might have had on state policy and the future of domestic commodity production in the area. Greater outbreaks of violent protests against the draggers, and the accompanying media coverage, might have influenced more significant government policies that favoured the fixed-gear fishery. As it was, the state did create some specific policies aimed at helping the fixed-gear fishery during this period. These policies, however, failed to provide much hope for most fixed-gear fishers.

5
The Fall, 1988–92

As the residents of northwest Newfoundland maintained their delicate balance between social conflict and social cohesion during the late 1980s and early 1990s, they were clearly waiting for the state to take a decisive position on the future of the fishery. As discussed in the previous chapter, many residents had formed opinions about the cause of the decline in the fishery and were asking the question, "Can the fishery be saved?" Obviously, part of the answer hinged on what was identified as the actual cause of the decline. Residents, however, were aware of another facet of the situation: the type of fishery desired by the state. Hence, many residents were also asking the question, "If the fishery is saved, who will it be saved for?" When no clear answer was forthcoming from the state, residents attempted to decipher the state's position from its specific policies. These policies, however, were often ambiguous in design and inconsequential in their effects. This left considerable room for a diversity of interpretations and opinions about both the cause of the cod stock decline and the state's plan for dealing with it.

Ecology of Gulf Cod and Its Decline

It would greatly increase our ability to evaluate the various opinions about the problems in the fishery discussed in Chapter 4 if we could compare them to what was *actually* causing the decline of the Gulf cod stock during the late 1980s and early 1990s. Although the current state of scientific knowledge about Gulf cod will now be summarized and several ecological factors that might have contributed to its decline will be discussed, this knowledge is imperfect and open to dispute. There is still considerable room for the social construction of knowledge, even among scientists.

The "Gulf cod" actually consist of two partially discrete stocks (Scott and Scott 1988:266). The fishers of northwest Newfoundland harvest cod primarily from the stock that migrates to the northern Gulf (areas 4R and 3Pn) between the Strait of Belle Isle in the summer and fall and then to the waters off Port aux Basques during the winter. A second stock migrates in the southern Gulf. This stock is of little direct importance to fishers from northwest Newfoundland, except that it may make up part of their catch during the "winter fishery." A more controversial issue concerns the possible mixing of the northern Gulf cod with "northern cod" (the stock found in fishing areas 2J, 3K and 3L of the northwest Atlantic) through the Strait of Belle Isle. Although "scientists have not quantified the extent" (Marshall 1990:3) of seasonal intermixing of these two stocks, "CAFSAC concluded that no major incursions of 2J3KL ["northern"] cod had taken place between 1983 and 1986, and although stock exchanges exist with 2J3KL in the Strait of Belle Isle, the 4RS, 3Pn cod stock appears relatively discrete" (Marshall 1990:23). This finding is highly disputed by many dragger fishers who insist that much of the fish they catch in the northern part of the Gulf during the late summer is actually northern cod. This insistence may be related to the fact that these skippers often blame foreign offshore trawlers, who heavily fish the northern cod (but not in the Gulf), for the decline of the Gulf cod stock.

One of the most hotly debated issues is the spawning period of cod, which ranges from February in the north to December in the south, depending on the particular stock in question (Scott and Scott 1988:267). Although there is little direct scientific evidence, it is widely held that dragging during the spawning period is harmful to the cod stocks. Hence, a frequent complaint against the winter fishery is that it is particularly destructive because it catches fish when they are spawning. The evidence actually suggests that most spawning among the 4R, 3Pn cod occurs slightly later in the spring when the cod have moved further north (Marshall 1990:21).

One of the reasons scientific knowledge about the cause of the decline in the Gulf cod stock is so tentative is the difficulty scientists have experienced even in determining if the stock is declining. The size of the Gulf cod biomass was assessed during the first years of quota allocations on the basis of two indices. The first was the trawler CPUE (catch per unit effort) that drew estimates directly from catches recorded by the dragger fleet. Fixed-gear fishers had always been skeptical of this method for several reasons. First, "under the table sales" meant that measures of the harvesting level that were used to compute stock size estimates were lower than the actual harvests. Hence, more of the stock was being harvested each year than the estimates of stock size allowed for. Second, catch records suggesting that large concentrations of cod were still available failed to adequately take into account the tremendous increases in the technological efficiency and skill levels of the dragger fleet. Hence, fish continued to appear abundant when in reality the fleet had just become better at catching the diminished amount of fish that were present. The skepticism over CPUE stock estimates spread to many DFO officials when "[t]he exact biomass value for 1989 could not be determined with precision, due to difficulties with divergent abundance indices" (Marshall 1990:20). While the CPUE indicated that the stock was "relatively stable and lightly exploited" (Marshall 1990:20), the winter survey, which formed the second method of assessment, indicated that the stock was "exploited above Fmax and declining at a rapid rate." As a result of such inconsistencies, fishery officials began to include other types of data, such as fixed-gear catches, in their estimates of the size of the Gulf cod stock and eventually concluded that the decline warranted a closure of the fishery (see Chapter 6). The difficulties in assessing the health of the stock, however, not only delayed responses to the problem of the declining resource, but they continue to plague plans for the future of the fishery.

Surveys have also indicated low recruitment in the Gulf for all age classes since the early 1980s except for the 1986–87 age class. This age class, therefore, forms much of the hope for the future of Gulf cod. It is no coincidence that the debate over regulatory measures peaked in the early 1990s as this age class approached both the age of reproductive maturity and the age when it could be caught.

Except when they are young (up to 50 centimetres in length), cod are bottom feeders capable of consuming a wide variety of prey including capelin, herring, redfish, other cod, hake, sand dollars, marine worms, mussels and combjellies, to name only some (Scott and Scott 1988:267). Cod prefer cool temperatures within a fairly wide range from -0.5°C to 10°C and are capable of adjusting quickly to transitions from warmer to colder water by generating internal "anti-freeze" within several days (Richard Haedrich, personal communication). Although government scientists remain convinced that environmental change has been significant in bringing about the recent collapse, evidence from the Atlantic area, at least, suggests that changes in water temperature and other environmental factors have not been sufficiently radical to produce such a consequence (Hutchings and Myers 1994; Haedrich and Fischer, forthcoming). Overfishing is by far the most significant factor in the northwest Atlantic and probably in the Gulf of St. Lawrence as well. Nevertheless, state managers and fishers remain unconvinced, and both have had to plan their strategies for dealing with the declining catches in a context of uncertain knowledge.

Hollow Victories for Fixed-gear Fishers

The 1988 fishing season opened with what appeared to be major new regulations aimed at restricting the dragger fleet and protecting the future of fixed-gear fishers. The most dramatic regulation was the imposition, during the winter fishery, of a buffer zone that excluded the draggers from waters less than 100 fathoms deep surrounding the town of Port aux Basques on the southwest coast. This "100-fathom edge," as it became known, was designed to save the southwest coast fixed-gear fishery by giving it exclusive access to the traditionally lucrative fishing grounds in the shallow water near Port aux Basques.

The dragger fleet felt this restriction would be devastating to catches and force them into dangerous trips offshore. They felt it marked a shift towards a clear anti-dragger policy on behalf of the DFO and they also pointed out that this policy would harm the residents of the southwest coast whom it was supposed to benefit because the economy of these communities was largely dependent on processing the fish caught by the draggers during the winter. Southwest coast residents responded to this argument by stating that the current situation only resulted from the fact that the local fixed-gear fishery had been destroyed by the draggers.

As they expected, the draggers did struggle to catch fish during the 1988 winter fishery. Although many dragger skippers blamed the poor catches on the 100-fathom edge restriction, and a number of boats were cited for violating this restriction (LeDrew 1988), the 100-fathom edge was probably only a small part of the problems faced by the draggers during 1988. Catches were also slowed by greater monitoring and surveillance, while marketing problems hindered the dragger fleet throughout the year. Indeed, the 100-fathom edge ceased to be an issue in following years because the cod failed to return to their traditional inshore waters and the dragger fleet managed to have continued success in the deeper waters outside the zone.

A second apparent victory for the fixed-gear association was the implementation of a buffer zone in the northern Gulf. This prohibited draggers from operating within two miles of shore from Point Riche to Cook's Harbour and along the northern shore of the Strait of Belle Isle. The primary purpose of this restriction was to protect cod traps and gill nets from being destroyed by draggers. It was also hoped that the fish would no longer be driven away from the nearshore fixed gear by the sound and disruption of the draggers. Although this measure did reduce the destruction of gear, it did little to save the fixed-gear fishery. Cod simply failed to come into the traditional inshore waters. Consequently, many cod trap fishers left their traps on shore between 1988 and 1991, as there was no point in even placing them in the water.

Although these new buffer zones were seen by many dragger fishers as clear signs that the state had turned against them, they actually did little to hurt the dragger fleet. The buffer zones also did little to produce more fish for the fixed-gear fishers. As a result, they did little to convince fixed-gear fishers that the state was committed to protecting their future. Many fixed-gear fishers, convinced that they had lost their traditional sources of income, were forced to search for other means of earning a living. For those who left the fishery and went to the mainland in search of work, the end of DCP was at hand if the migration proved to be permanent (House et al. 1989). Others, however, turned to new patterns of fishing in attempts to survive.

The problem with turning to alternative fisheries was simply that there were few other species to replace the lost Gulf cod. Even though lobster came to dominate cod in the late 1980s, this only helped fixed-gear fishers who held a lobster licence. And, even those who had lobster licences could only rely on three to six weeks of insurable income from this fishery. Lump roe might provide an additional two or three weeks, but this fishery was unreliable and often overlapped with the lobster season. Capelin also interfered with lobster fishing and continued to be unpredictable.

Fishers with longliners and scallop licences had the option of pursuing the scallop fishery, which was often known to produce sufficient weeks of income to qualify for UIC. Scallop prices, however, dropped to three dollars per pound during 1989, making what was generally considered the most gruelling and uncomfortable fishery in the area even less attractive. Given the absence of alternative species in the Gulf, the solution for many longliner and fixed-gear fishers was to leave the Gulf.

The problems experienced by both the fixed-gear and mobile-gear fishers during the early 1990s also produced the search for "underutilized" species. Hence, several dragger skippers, who were temporarily without a vessel suitable for otter trawling, combined with fixed-gear fishers who were unable to earn sufficient insurable weeks of income from cod fishing to pursue deepwater gill netting of turbot off the northern Labrador coast. Although several of these vessels met with moderate success, there was already talk about today's underutilized species becoming the endangered species of tomorrow.

The Labrador Migration

Newfoundland fishers (especially those located from Conception Bay to White Bay) have been travelling to harvest the late summer influx of cod along the coast of Labrador for nearly two centuries. Traditionally, however, the practice had been limited on the northwest coast to a short migration across the Strait of Belle Isle (Palmer 1995). After one or two weeks of fishing near their home communities, fishers would shift their cod traps to the Labrador side of the Strait where they would stay in tents, or with friends and relatives who lived there permanently. After one or two weeks of additional cod-trap fishing, they would attempt to jig for another week or two in the Strait.

The first increase in the Labrador migration from the northwest coast occurred with the advent of the longliners, because the longliners, which also served as housing, made further and longer-term migrations practical. Longliners could also be used to fish gill nets further from shore. A few longliner crews fished 50–200 miles further north along the Labrador coast as early as the 1970s. This migration increased in the mid-1980s when more longliners began to choose the journey to Labrador over the depleted stocks and the danger of losing gill nets to draggers in the Gulf.

The Labrador migration still remained a minor part of the northwest Newfoundland fishery until large numbers of small-boat fishers joined longliner fishers in the late summer migration of 1988. Small-boat fishers had either to make the journey on their own or load their boats and materials onto the decks of collector boats, which had begun to follow the fishers in order to purchase their fish and return it to Newfoundland for processing. Once there, the small-boat fishers built their plywood camps and formed small shanty towns, primarily in and around the community of Black Tickle.

The Labrador migration boomed during 1989 and was a temporary economic saviour for small-boat fishers. The abundance of fish attracted an estimated 7,000 fishers from Newfoundland. More than 12 million pounds of cod and nearly one million pounds of turbot were purchased in the Black Tickle area during the late summer and fall of 1989 (Palmer 1992). Indeed, the main problem—besides a gear conflict due to the overcrowding of cod traps and gill nets—was that the catch far exceeded the capacities of the few local Labrador fish plants and the handful of collector boats that brought fish back to the island of Newfoundland for processing.

Table 9. Cod and Turbot Purchased by Black Tickle Area Fish Plants and Collector Boats from Other Areas, 1988–90 (kg)

	1988	*1989*	*1990*
Cod			
Black Tickle Area Plants	5,548,275 (83%)	5,551,478 (46%)	2,737,946 (42%)
Collector Boats	1,140,070 (17%)	6,503,753 (54%)	3,791,018 (58%)
Total	6,688,345	12,055,231	6,528,964
Turbot			
Black Tickle Area Plants	145,863 (30%)	31,663 (4%)	4,577 (1%)
Collector	348,177 (70%)	803,930 (96%)	539,733 (99%)
Total	494,040	835,593	544,310

Source: DFO, Moncton.

In a situation with many parallels to the winter fishery on the southwest coast, the influx of fishers from other areas caused considerable ill-feeling among local residents of Labrador. Not only had Labrador fishers to compete with outsiders on the overcrowded fishing grounds, but local fish plant workers also had to contend with the collector boats in order to process the fish (see Palmer 1992). Table 9 indicates that, during the glut of 1989, Black Tickle area fish plants were still able to operate at peak capacity, but the success of some collector boats in the competition for the scarce supplies of fish in 1990 meant that many of the local plant workers failed to attain even half the number of insurable weeks needed to qualify for UIC.

The DFO's proposed solution to the Labrador conflict also bore striking similarities to the 100-fathom-edge regulation that had been implemented to solve the winter fishery conflict. A buffer zone that excluded the vast majority of outsiders from the shoal water immediately surrounding Black Tickle was implemented for the 1991 season. Like the 100-fathom edge, this buffer zone was very unpopular with most of the excluded fishers. Its impact on the fishery, however, could not be calculated because fish failed to migrate to this area during 1991. This also meant that an additional policy, based on the principle of limiting access to only those fishers who had fished in a minimum number of previous seasons, could not be implemented.

Small Cod and the Pacification of Fixed-gear Fishers

Given that the "last resort" (*Northern Pen,* July 27, 1990) Labrador migration failed completely in 1991, this year might have been expected to be a crisis point that would lead to violent protests by fixed-gear fishers. The reason it was not was the unexpected arrival of cod in traps along the west coast of Newfoundland during the early and middle summer of 1991. Fishers who had left their cod traps in their stores for several summers loaded them in their pickup trucks and followed the fish for over a month, some travelling as far south as Green Point in June and as far north as Cook's Harbour by late July. Although some crews were still planning to migrate to Black Tickle later in the summer, many no longer felt the need since they had acquired sufficient fish closer to home.

This "unexpected bonanza" (*Northern Pen,* July 19, 1991) in the form of the return of cod to cod traps along the northwest coast did more than just diminish the significance of the failure of the Labrador fishery; it also quieted the protests of fixed-gear fishers. Fishers, who had already lost much of their ability to organize protest with the near collapse of the Coalition for Fishery Survival (which had grown from the Fixed-gear Association in 1989), were pacified by the presence of fish in their nets. As one wife of a fixed-gear fisher said, "This is the first time in years anyone could even think about buying a new car." This pacification occurred despite the fact that catches were still only a shadow of previous levels, and fishers were now forced into being "mobile fixed-gear fishers" in order to catch sufficient amounts. The most disturbing fact of all, however, was the small size of the fish. Fishers were aware that this was a sign of an unhealthy stock, especially since most realized that they were catching the 1986–87 year class that was thought to be the only healthy year class in the Gulf, and they were catching it before it had reached reproductive maturity.

Still, the presence of a cod-trap fishery, after several years of leaving their traps on shore, took much of the urgency out of the fixed-gear fishers' demands for changes. The appearance of variable amounts of cod in cod traps along the coast in 1992 and 1993 continued to dampen their protests. Indeed, by the spring of 1992, a role reversal was about to take place. While the Gulf cod stock appeared to be improving, dire rumours were beginning to spread about the northern cod (CBC fisheries broadcast, February 19, 1992).

The Draggers: 1988–92

Although they remained a wealthy elite in comparison to most fixed-gear fishers, dragger skippers and crew members faced difficult economic times during this period. One DFO official even speculated that many of the skippers would never have gone into the dragger fleet in 1975 if they had known what their 1992 quota would be. In any case, they were no longer envied by fixed-gear fishers to the extent that they had been in the past. As one fixed-gear fisher stated, "Now I wouldn't even want a dragger, I'd like to have a dragger quota [to lease to someone else], but I'd hate to have to make payments on a dragger." Some skippers talked about wanting state buy-back programs to help them out of the mobile fishery, and a DFO official estimated that 60 percent to 70 percent of the fleet would switch to fixed-gear if they were provided "decent compensation." Given that "decent compensation" was unlikely, the general mood among the dragger fishers was one of "cutting back" and "holding on." For example, when one dragger skipper was complimented on the spaciousness of his home, he replied "Now I wish I had built it twenty feet by twenty feet." Other dragger fishers were busy installing woodstoves into the homes they had built during the mid-1980s with only "modern" electric heat. Significant "downsizing," however, was difficult for draggers because there was little chance of finding a buyer for the homes they had built during the glory days of the fishery. The economic dreams that had started the dragger fleet were not dead, but they had been tempered by new economic realities.

Although it appeared to lack a clear view of the long-term future of the dragger fishery, one unavoidable fact faced the state in 1988: fleet reduction. What had been a tactic of "last resort" in 1982 and a hoped for by-product in 1984 was now a necessity. There were not only too many vessels for the shrinking quotas, but the average operating costs and harvesting capacities of these vessels continued to grow to the point where one DFO official stated that comparing the draggers of the 1970s to the 65-foot draggers of the 1990s was like comparing "pea-shooters to bazookas." Hence, the only question was how this reduction would be accomplished. Although the most effective method might have been to return to a competitive fishery regulated by a fleet quota, this would have been seen as "ruthless" by many skippers and could have led to considerable public protest. One dragger crew member stated that the reduction of the fleet by a fleet quota would have been a "very callous way, and they [the DFO] knew it wouldn't have been very popular. They knew they were going to have hell to pay if they did it that way." Returning to a fleet quota would also have been seen as saying that the enterprise allocation system had been a failure. Instead, it was decided to increase the transferability of the enterprise allocations and transform them into a limited individual transferable quota (ITQ) system.

ITQS

Although it is often hoped that ITQs will have beneficial conservation as well as positive economic effects (Crothers 1988; Dewees 1989; Pearse 1980), the main theoretical goal of transforming enterprise allocations into ITQs is to rationalize fisheries by increasing the power of market conditions to influence supply and demand (see Binkley 1989; Crothers 1988; Boyd and Dewees 1992; Libecap 1989; Eythorsson 1996; McCay and Creed 1990). Less viable enterprises would be able to leave the fishery by selling their quotas and licences to more viable enterprises. Hence, over time, a fleet reduced in numbers but with increased profit margins per enterprise would develop. Although some limited forms of temporary transferability were introduced into the second three-year enterprise allocation program (1987–89), transferability was greatly expanded as part of a new program to run from 1990 to 2000.

ITQ systems are also thought to have additional benefits for the state. First, they could replace many of the regulations concerning entrance to the fishery and hence reduce the role of the state in regulating the fishery (for discussion of such co-management, see Davis 1984; Dewar 1990; McCay and Acheson 1987;

Taylor 1987; Ostrom 1990, 1992; Singleton and Taylor 1992; Jentoft 1989; Pinkerton 1989). Second, the state could receive considerable rent from an ITQ system by charging substantial fees for licences (Boyd and Dewees 1992). In northwest Newfoundland, however, the first of these effects failed to occur because the existing regulations remained. The attempt to invoke additional fees for licences was also met with considerable resistance.

The actual motivations for implementing a management policy in a specific setting may diverge from the general theoretical goals of that type of management system. In the case of the northwest Newfoundland draggers, several fishery officers and fishers speculated that one of the reasons for the adoption of the ITQ system was to help ensure that the Newfoundland region's overall quota would be caught each year. This was important because it was feared that the failure of the region to catch its quota might lead to the redistribution of the uncaught portion to other regions in Atlantic Canada during the following years (Copes 1986). At the beginning of the enterprise allocation program, this problem was handled by skippers being asked to report voluntarily their inability to catch the quota so that it could be reallocated to other vessels in the fleet before the end of the summer. Skippers were reluctant to make such reports, however, out of fear that their individual allocation would be reduced the following year. Hence the voluntary reporting system was replaced by a program that would automatically redistribute all uncaught allocations to the entire fleet on the fifteenth of August. This regulation, which was unpopular with the smaller boat operators who had more difficulty catching their quota quickly, especially if they experienced breakdowns early in the year, stayed in effect until the ITQ system produced an alternative means of maximizing the fleet's catches.

One obvious drawback to an ITQ system is the decrease in employment that accompanies fleet reduction (Crowley and Palsson 1992). Since this would be socially disastrous in an area like the Great Northern Peninsula, the ability to accumulate quotas was limited to the equivalent of two 65-foot quotas. This fact, along with the relatively small number of enterprises in the fleet, limited the ability of market forces to operate (Boyd and Dewees 1992).

ITQ systems are also thought to increase the skippers' perception of themselves as entrepreneurial businessmen. In northwest Newfoundland, this did generally occur among the dragger skippers, but it did not erase the previously existing egalitarian ethic. The fact that many northwest Newfoundland dragger skippers consider themselves businessmen is quite evident in the following dragger skipper's statement: "I've worked hard to build up a business. I've invested half a million dollars in it, and I'm damn sick and tired of outsiders coming in and trying to ruin it." Some of the most successful skippers have also invested their profits in non-fishing businesses. One skipper, however, still clearly expressed egalitarian values when he objected to the new ITQ system on the grounds that "ITQs just make the elite more elite." The opinions of skippers toward transferability at the time of our 1994 survey also indicated that their transformation into capitalist businessmen had not been complete. Although the vast majority of skippers (98 percent) had come to favour the enterprise allocation system, 70.5 percent of those interviewed in 1994 (see Chapter 6 and Appendix) were against transferability. Except for a few of the skippers of the largest vessels, dragger skippers did not like the idea of a few "quota kings" developing and dominating the fishery by driving other skippers out (see also Dewees 1989). Comments during interviews indicated that, in addition to fearing that they would be one of those forced out of the fishery, the accumulation of quotas by driving others out was seen to go against the previously described egalitarian tradition. Indeed, even the ability to accumulate two 65-foot quotas was seen as threatening by many skippers.

The weight of opinion against the accumulation of quotas may have been one reason few transfers occurred between 1990 and 1993. By the end of this period, there were 93 licences left out of the original 115. Six or seven of these were withdrawn by the DFO for violating time restrictions on inactive periods. The others were split among brothers or consolidated onto a single boat. During 1993, only seventy-six of the ninety-three licence holders actually had boats that were fishing. These seventy-six boats were fishing the quotas from the other eighteen licence holders on temporary transfers.

The condition of the fishery at this time also contributed to the low amount of combining. On the one hand, the fishery was in a state of decline and its future appeared precarious. Hence, only a few of the most optimistic and enterprising skippers were willing to invest $50,000 to $150,000 for an additional licence and quota. On the other hand, the dragger fishery remained the most lucrative economic option available to most skippers, and if it should survive the current crisis it would continue to be so in the future. Hence, relatively few skippers were willing to take the drastic step of selling their licence and abandoning the dragger fishery permanently. If a skipper lost a boat or missed a season due to repairs, it was much safer just to lease their quota temporarily to another fisher, while retaining their licence. Such temporary leasing was also economically viable because the skipper could receive about 25 percent of the value of the fish, and sometimes an additional 12 percent if he worked on the boat to which he leased his fish.

As a result of these factors, the ITQ program met with only limited success in the goal of reducing the dragger fleet. Although a reduction of approximately 20 percent occurred between 1990 and 1993, it was generally felt that most of the vessels that left the fishery would have done so with or without the ITQ system. Some skippers speculated that the ITQ system actually kept a number of skippers in the fishery who would have otherwise left. One dragger crew member speculated that the ITQ system "gave everybody the same chance . . . [and] it gives a person who really didn't work that hard a better chance to stay into it." In any case, it was generally considered by the DFO and many dragger skippers (see Chapter 6) that the fleet was still much too large in 1994 and needed to be reduced to sixty-five vessels or less.

Dragger Skippers Face a Declining Resource

Even though the majority of skippers hung on to their enterprises during this period, economic pressures continued to grow for them. The major problem was the continuing decrease in their cod quotas, the effect of which was magnified by significant increases in surveillance at sea and the dockside monitoring of landings. This greatly reduced the amount of unreported sales; charges were laid against a number of skippers and several boats were temporarily seized. Falling cod quotas also intensified the need for shrimp. Although for several years there had been rumours of new shrimp licences being awarded to non-shrimpers, they had never materialized. By 1988, shrimp had become equally or more profitable than cod, and the forty-eight non-shrimpers formed a lobby group to attempt to obtain shrimp licences. Although rumours of licences for the entire fleet persisted, few new licences ever materialized. This was possibly the result of anxiety over the health of the shrimp stock. Certainly the destructive by-catch of the shrimp fishery was a very real concern, but most dragger skippers also realized that the allocation of new shrimp licences to help enterprises survive went against the basic goal of reducing the fleet. To pacify the non-shrimpers, plans were implemented to give them greater access to redfish stocks, but this did little to offset the disparity between shrimpers and non-shrimpers.

Despite their own hardships and feelings of unfair treatment at the hands of the DFO, by 1990 there was growing recognition among dragger skippers that the stock appeared to be in danger and growing acceptance of reduced quotas. Although fishing remained relatively good during the winter outside the 100-fathom edge, daily catches in the rest of the Gulf had dropped into the 2,000–7,000 pound range. The size of the fish being caught was now so small that almost all skippers were aware there was a problem. Asked if the dragger skippers were concerned when they first noticed the fish getting smaller in 1986, a skipper answered no, "But as the years went by, by 1990 in particular, people were really noticing that the fish was real small and the catch rates were low and there wasn't many marks. You couldn't find fish either. You'd look the normal places for finding fish and it wasn't there."

This realization that there was a problem with the stock combined with the implementation of dockside monitoring and greater surveillance at sea produced a major decline in illegal fishing practices. Although rumours of illegal practices continued, since even the skippers themselves could never be sure if anyone in the fleet was cheating, Palmer observed no instances of liner use, excessive dumping or misreported sales during extended participation in the fishery through the 1990, 1991 and 1992 seasons.

Along with greater concern for the stocks came an even greater demand by the dragger skippers for increased surveillance. The skippers were, in effect, pleading with the DFO to force them to stop their own profitable but destructive fishing habits. The reason for this apparent paradox was, once again, the view that one was forced to cheat whenever anyone else in the fleet was profiting through illegal fishing activities. Hence, anything short of 100 percent monitoring coverage was seen as unsatisfactory because it allowed suspicions that some boats were cheating to continue. When the DFO suggested partial onboard monitoring, fishers complained that some of the known violators would end up not being monitored. Spontaneous boardings at sea were also claimed to select certain vessels while avoiding others. By November 1990, some skippers were even suggesting that they might start reporting violators to the DFO themselves, but we are unaware of any actual occurrences. What did occur was a growing frequency of dragger skippers admitting to past illegal fishing techniques in an attempt to dissuade their peers from continuing such practices. One observer at a meeting of dragger skippers and DFO representatives during the autumn of 1990 even commented that it "was getting to be like an AA [Alcoholics Anonymous] meeting with everyone standing up and saying 'I'm [so and so] and I've been using liners, misreporting and dumping fish at sea. Please [DFO] help me stop."

The mood of the dragger fleet began to change at the beginning of 1991 as a result of three factors. First, the fleet was hit with quota reductions of approximately 27 percent from the previous year. Although a reduction was expected, the amount was seen as severe. One skipper stated "They [the DFO] didn't do anything to save the fishery for years. Now they go hog wild!" Some skippers saw the reductions as a clear sign that the state had decided to get rid of the dragger fleet altogether. Most, however, saw it as a means of reducing the fleet: "The boats with only one quota won't have a chance. Everybody will have to double up to survive." Not surprisingly, fixed-gear fishers viewed the plight of the draggers with little sympathy. Some stated that the quota reductions were "a good first start," and another fixed-gear fisher simply said, "Now they know how it feels."

The second factor that has influenced changes in the attitudes of dragger skippers concerns a common problem with enterprise allocation and ITQ systems of management: the enforcement often turns out to be expensive (Crothers 1988; Dewees 1989; Boyd and Dewees 1992). Even providing 50 percent monitoring coverage for the upcoming winter fishery was going to be so expensive that the DFO attempted to impose "user fees" that ranged from $1,425 to $3,415 per vessel depending on the size of the vessel and whether or not the owner had a shrimp licence. This requirement was met with strong opposition from the dragger fleet and the union. In fact, the fleet decided to refuse to pay the fees and remained tied up during the first few weeks of the scheduled winter fishery. Although the debate was often couched in purely economic terms, the imposition of fees was seen as having greater implications. First, it was thought to be a further sign that the DFO had abandoned the dragger fleet and was trying to force many, if not all, vessels out of the fishery. Having to pay for their own regulation was also thought to place too much of the blame on the dragger skippers. Many skippers had begun to admit there was a serious problem with the Gulf cod stock and an increasing number had begun to accept partial blame. Having to pay the DFO to stop the illegal fishing practices of some draggers was, however, seen as admitting much more guilt than they felt was deserved. Because they had to pay user fees while the fixed-gear fishers did not, they felt they would be perceived as being punished for having singlehandedly destroyed the stock. This they refused to accept. After all, hadn't the DFO's lack of enforcement in the past been the real cause of the illegal practices?

The third factor aggravating the situation was reports of surveys indicating that the Gulf cod stock might be larger than previously assumed (CBC fisheries broadcast, February 4, 1991). This reinforced the dragger fishers' growing conviction that the quota cuts were just a means to reduce or eliminate the fleet, and the user fees were just part of the plan. This helped to solidify the entire fleet's will to refuse to fish until user fees were reduced. It even appeared that the FFAW was once again a powerful force, as several union meetings were held and the skippers voted to remain tied up. This image of renewed solidarity dissolved, however, when the CBC fisheries broadcast of February 8, 1991 reported that the fleet would begin fishing and pay the user fees "under protest." The dragger skippers, who had voted just the previous day to remain tied up at least until another meeting could be held in a few days, were startled. A few were even angered by the fact that their vote had been ignored, but most simply saw it as "typical" and prepared to go fishing. Hence, the winter fishery finally began, but the attitude of many dragger skippers had changed. The events at the beginning of 1991 had caused a retrenchment in the attitudes of many skippers. Fewer dragger skippers were now willing to accept significant responsibility for the collapse of the stock, and fewer were willing even to admit there was a problem with the Gulf cod stock. By the autumn of 1991 some skippers were even hoping for increased quotas in 1992.

Fish Plants

The declining quotas and catches caused a corresponding decline in fish plant operations along the northwest coast. By 1989, person-year employment in area fish plants had dropped to about 2,400, a 40 percent decline from 1984 (Marshall 1990). The only reason this amount of work had been maintained was that local plants, which were once providers of fish to other areas of the province, were now importing large quantities of fish. By 1990 nearly a dozen fish plants in the area relied on approximately ten million pounds of fish brought by trucks from the winter fishery and other parts of the island and by collector boats from the Labrador fishery. This obviously increased tensions with the residents of these other areas who were also in need of fish to process (Palmer 1992).

As fish plant jobs became scarcer, they became even more precious. This meant that the loyalty of draggers to their hometowns was even more crucial, because such loyalty could now determine whether or not the community's fish plant stayed in operation. At the same time, however, dwindling quotas meant there was more incentive for skippers to sell their fish to whomever offered the highest price. Ironically, this issue was perhaps most crucial for the wives of fixed-gear fishers whose husbands' fishing incomes had become minimal. As one fixed-gear fisher stated after a poor day of lobster fishing in 1991, "Sure am glad the wife's working [in a fish plant] today." The uncomfortable nature of this situation helps explain the euphoria experienced when small amounts of cod showed up in cod traps during 1991 and 1992. Despite the small size and limited amounts of this fish, these catches temporarily allowed fixed-gear fishers to regain their roles as economic providers to their households.

The Role of the State

The ambiguity of state policy towards the dragger fleet reached its apex during the period from 1988 to 1992. Many fixed-gear fishers and nonfishing residents were convinced that the state backed the draggers and saw them as the sole fishery of the future. As evidence of government support for the draggers, residents pointed to such events as the conflict during the 1991 winter fishery. This involved the DFO's closure of the fishery due to large by-catches of cod in the redfish fishery after the cod quota had been exhausted. The dragger fishers responded by blockading the Port aux Basques to North Sydney ferry and staging a violent occupation of the DFO office in Port aux Basques (*Maclean's* 1991; Palmer 1992). These actions were followed by the reopening of the fishery three days later. Although the DFO emphasized that the reopening had not been the result of the protests, many fixed-gear fishers saw this as another example of dragger fishers getting whatever they wanted.

Dragger fishers, however, did not share this view of the government's position; many felt abandoned and betrayed by the state during this period. They pointed to the new buffer zones as evidence that they were being restricted in order to ensure the future of the fixed-gear fishery. Although they had always blamed the lack of enforcement for causing the illegal fishing practices, they now saw the imposition of user fees for increased enforcement as placing an unfair share of the blame on the dragger fleet. They saw the incident in the 1991 winter fishery not as a demonstration of their own power but as a clear indication that the state was trying to prevent enterprises from surviving. Of course, the most convincing evidence of lack of state support was the rapidly dwindling quotas. Although many skippers were willing to admit some need to reduce quotas during the late 1980s, they became more skeptical of this need when they felt their own existence being threatened in the early 1990s.

The extreme differences in opinion about the role of the state during this period resulted from more than differing self-interests. State policy at this time really was ambiguous. On the one hand, the fixed-gear fishers were right in assuming that there was still a general commitment to dragger technology as the fishery of the future. On the other hand, however, state officials were reluctant to discuss what that meant for the fixed-gear fishery and much of the population of the region. Many dragger fishers were also correct in their view that the future dragger fishery envisioned did not necessarily include their particular enterprise. As Sinclair observed,"[a]lthough they do not say so publicly, interviews with senior fisheries managers indicate that they see no future for the small open boats of the traditional fishery . . . [and] [t]hey also feel there are too many larger vessels" (1990:40). Many local fishers also realized the nature of this situation. As one skipper of a small dragger stated, "You know, the real division isn't between the draggers and the fixed-gear, it's between the big [successful] draggers and everyone else on the peninsula." The growing realization of this division was also reflected in one skipper's suggestion that the larger draggers had secretly been in support of the user fees during the 1991 tie-up because they might have forced some of the smaller, more marginal enterprises out of the fishery.

Conclusion

The period between 1988 and 1992 saw a crisis in the Gulf fishery that pressured the state towards taking a decisive stand on the dragger fleet. Although a general position in favour of maintaining part of the fleet for the future can be discerned during this period, events conspired to keep the state's position largely implicit. The need to reduce the dragger fleet was partially postponed because of increased catches of redfish. The ITQ system, as implemented, also failed to force large numbers of boats out of the fishery and may have kept some vessels in the fishery that might otherwise have left. The state might have been forced into a clearer position by the growing complaints from the fixed-gear fishery and calls from both the Coalition for the Survival of the Fishery and the FFAW for drastic measures, including a moratorium with compensation. These demands were growing so loud by 1990 that they could hardly be ignored; however, they were muted by the return of moderate amounts of cod-trap fish in 1991. This left many northwest Newfoundland residents asking the same question a longliner skipper asked on an October 22, 1990 CBC special entitled "The Fishery of the '90's": "Is the government intentionally starving us, or is it just stupidity?"

Cod trap crew during cod trap revival

Redfish from the winter fishery

6
The Closure, 1993–95

During the summer of 1990 some of the cod-trap crews in the St. John's area of Newfoundland were enjoying their best season ever. For these crews the fabled abundance of the northern cod had never been more real. Meanwhile the Gulf cod stock was being declared the greatest ecological disaster in Atlantic Canada (*Northern Pen,* September 19 and 26, 1990) and calls were made by the Coalition for Fishery Survival, the FFAW and various political leaders for a moratorium on cod fishing in the Gulf. Increasingly, fixed-gear fishers and others demanded the abolition of the local dragger fleet. By the summer of 1992, however, the situation, or at least the talk about the situation, had changed. The Gulf cod fishery had been proclaimed "on the road to recovery," (CBC fisheries broadcast, April 13, 1992) while a moratorium was imposed on the northern cod. These rapid changes in the perception of the Gulf cod stock and its health relative to the northern cod were part of the reason for the diversity of opinion among northwest Newfoundland residents about the closure of the Gulf cod fishery. It is necessary to understand these complex and rapidly changing circumstances surrounding the 1994 Gulf cod moratorium in order to understand the role of the state in determining the fate of DCP in northwest Newfoundland.

The Impact of the Northern Cod Moratorium

It is impossible to comprehend the Gulf cod moratorium at the beginning of 1994 without grasping the impact of the northern cod moratorium that began during the summer of 1992. The first effect of the northern cod moratorium was to increase the skepticism that had always existed towards scientific data concerning the health of cod stocks. The estimates of northern cod during the late 1980s now appeared to have been grossly inflated because of misreporting by fishers and inadequate survey methodology (Steele et al. 1992; Finlayson 1994). This revelation might have made northwest coast fishers leery of the 1992 survey showing a slight increase in the Gulf cod stock. Instead, many dragger skippers chose to see the inadequacies of the northern cod surveys as a reason to question the earlier pessimistic figures that had indicated a collapse of Gulf cod. They took this position despite being aware that Gulf cod estimates during the 1980s were based largely on the dragger fleet's catches and were likely to have provided overly positive assessments (see Chapter 5). Ignoring these facts continued the trend toward denial of any serious problem, a trend that had begun with the resistance to user fees before the 1991 winter fishery.

Fixed-gear fishers considered the northern cod moratorium unfair. More precisely, the fact that fishers in other parts of the island who relied on northern cod were receiving a compensation package was seen as unfair. Northwest Newfoundland fishers complained that their fishery had been effectively closed by the lack of fish for years during the period when some northern cod fishers were enjoying record catches, but they had never received compensation. These complaints, however, failed to grow into outraged demands for closure of the Gulf fishery and a similar compensation package, because of the moderate amounts of cod-trap fish that continued to be caught along the northwest coast during the summers of 1992 and 1993.

The presence of cod-trap fish along northwest Newfoundland also dampened demands for the abolishment of the dragger fleet. While there was still considerable talk among fixed-gear fishers to the effect that a cod fishing moratorium would be the perfect time to phase out the dragger fleet, much less effort was directed towards this end in 1994 than in 1990. Indeed, while fixed-gear fishers were certainly more concerned about the state of the stock and more in favour of the moratorium than their mobile-gear counterparts, their position was not emphatic. As one fixed-gear fisher stated in 1994: "I don't understand it. For years we weren't catching a fish and they left it [the Gulf cod fishery] open. Then last year we caught more fish out here [in our cod trap] than we have since I was a kid, and what do they do? They close it!" As this statement indicates, fixed-gear fishers felt the need for a closure, especially if there was to be suitable compensation, but the timing did not seem to make sense.

Because of collector boats that brought several million pounds of northern cod from Labrador to the Great Northern Peninsula for processing, workers at several fish plants qualified for the northern cod moratorium compensation package. This only increased the cries of injustice from workers at fish plants who did not qualify for compensation, despite the fact that they had little or no work. When the Gulf cod moratorium led to the closure of many area fish plants, workers scrambled to qualify for their own compensation packages. A few plants, however, did manage to stay open during 1994 by processing shrimp, scallops and lobster.

The Gulf Cod Moratorium: The Dragger Fishery

Although the Gulf cod moratorium did not go into effect until January 1, 1994, many participants in the dragger fisheries felt the economic blow during the summer of 1993 when the DFO "temporarily" closed the cod fishery because of a high percentage of undersized fish. When the fishery failed to reopen, dragger fishers lost the chance to catch the remaining part of their 1993 quota. Not only did this put a number of enterprises under financial pressure, many skippers did not believe the DFO's stated reasons for closing the fishery:

> They [the DFO] said it [the cod] was too small, [but] they just didn't want no big lot of fish coming in. . . . They had already made up their minds that there was no fish in the Gulf . . . [and] they didn't want too much coming in because they were afraid it would contradict what they were saying. They just wanted to close the fishery.

This skepticism over the DFO's stated reasons and the government's refusal to compensate the draggers or their uncaught fish (and only to provide support for vessel upkeep) increased the conviction that the closure had just been an excuse to force them out of the fishery.

Although the official moratorium of 1994 did not come as a surprise to the dragger fleet, our 1994 survey revealed considerable variation among the attitudes of dragger skippers towards the closure. First, in regard to the timing of the closure, Table 10 shows that 27.6 percent accepted that the timing was correct, and 29.3 percent even stated that the moratorium should have been declared earlier than it was. The following statement is typical of the opinions of the slight majority who agreed with the closure:

> It's a good move in a way, because I think it might have a chance of protecting the bit that's left out there, hopefully. That is, if we're gonna have rural Newfoundland. If we're looking at the long term, if we're gonna survive as rural communities, the fishery is the only thing we've got.

A sizable minority of 43.1 percent, however, stated that there should not have been a closure. The following comment is typical of those who held this position: "It should have been left open. The fish was small, but there was lots of it. I think they closed it just to get some of us [dragger skippers] out [of the fishery]. The fish was just an excuse."

Given the different opinions about the need for a closure, it is not surprising to find a similar division in regard to the perceived health of the Gulf cod stock and the desired length of the closure. Although most dragger skippers see a problem with the condition of the Gulf cod stock (see Table 11), a majority of 55.9 percent viewed it as only slightly worse than before, and only one-third (33.9 percent) stated that it was much worse. Table 12 shows the expected link between perception of stock condition and attitude towards the closure. The worse condition a skipper believed the stock was in, the more likely the skipper

Table 10. Attitude Towards Moratorium by Skippers' Vessel Size

	View of Timing of State Action (%)			
Vessel size	*Too late*	*Proper Time*	*Should Not Have Been Closed*	*Number*
Up to 55 feet	28.1	28.1	43.8	32
Over 55 feet	30.8	26.9	42.3	26
Total	29.3	27.6	43.1	58

Chi-square p. = .976, V=.03.

Table 11. Attitude Towards Current Stock Condition by Skippers' Vessel Size

	View of Current Stock Condition (%)			
Vessel size	*Same as Before*	*Slightly Worse*	*Much Worse or Extinct*	*Number*
Up to 55 feet	9.1	57.6	33.3	33
Over 55 feet	30.8	53.8	34.6	26
Total	10.2	55.9	33.9	59

Chi-square p. = .938, V=.05.

Table 12. Preference on Timing of Closure by View of the Stock

	Preference on Closure Timing (%)			
View of Current Stock Status	*Earlier*	*Status Quo*	*No Closure*	*Number*
Same as before	0.0	0.0	100.0	6
Slightly worse now	20.6	26.5	52.9	34
Much worse or extinct	55.0	35.0	10.0	20
Total	30.0	26.7	40.3	60

Chi-square p. < .001, V= .4.

Table 13. Preferred Length of Moratorium by Skippers' Vessel Size

	Preferred Length for Moratorium (%)			
Vessel Size	*1–2 Years*	*3–5 Years*	*More than 5 Years*	*Number*
Up to 55 feet	40.6	37.5	21.9	32
Over 55 feet	36.0	40.0	24.0	25
Total	38.6	38.6	22.8	57

Chi-square NS, V= .05.

was to consider that the moratorium should have been declared earlier (V=.40; Chi-square p.>.001). Not surprisingly, there were also similar differences in preferred length of the closure with 38.6 percent stating that it should not exceed two years, while 22.8 percent advocated a period greater than five years (see Table 13).

The dragger skippers were much more united in defence of their own fishery. The most frequent reasons they gave for the decline in the Gulf cod stock were seals (22 percent) and cold water (11.9 percent), with only five skippers (8.5 percent) stating that their own fleet was the major contributor. Similarly, 90 percent stated that draggers should be allowed to operate once the moratorium is over. Indeed, 80 percent intend to return to the mobile cod fishery when it reopens, and most of the remainder, the older skippers, want to retire and pass their licences on to their sons. In the event that draggers are not allowed in the future, only 47.1 percent would continue to fish by other means or for other species. Most would expect a generous buyback scheme.

In defence of their fishing technology, dragger skippers often stressed the importance of recent technical innovations in otter trawling. The adoption of square mesh and new kinds of lines on their nets were considered to significantly reduce the catch of undersized cod. Of even greater importance was the adoption of the Nordmore grate in the shrimp fishery. This grate separates the by-catch of small fish, which are released from the net during a tow, from the shrimp, which continue into the cod end. Dragger skippers had resisted this device for several years out of fear that it would cause the loss of large amounts of shrimp. After it was praised by some of the skippers who experimented with it during 1993, however, it won wide acceptance. These changes were claimed by many skippers to greatly reduce or eliminate whatever damaging effects the dragger technology might previously have had on the fishery. One skipper even stated: "With the Nordmore grate and all, the dragger fishery is now the most environmental friendly fishery on the go."

While nearly everyone agrees that the Nordmore grate and other technological changes have lessened the ecological damage caused by otter-trawl fishing, few residents who are unconnected with the dragger fishery feel it is now environmentally friendly. Many residents feel that even if the draggers do no ecological damage other than catch legal-size fish, the mere fact of their enormous catching capacity still makes them a threat to the future of the fish stocks. One dragger skipper who recently left the fishery described the situation this way:

> I think the dragger fishery can be a destructive fishery. I don't know if it's possible to be anything else but destructive. I'm not sure what damage it's doing out there to the bottom or what damage it's doing to the spawning grounds. In terms of catching fish, it can be destructive. We've seen that happen. . . . Even if we don't know the kind of damage it's doing to the bottom, obviously it's doing some. Even if it doesn't do any damage at all to the bottom, or shaking up the ecology out there, it still has to be looked at from the point of view of how it's used. It has to be controlled. It can't go on like it's been going on.

Large Draggers vs. Small Draggers

Given the growing concern that the state's long-term goal was a fishery based on some, but not all, of the current draggers, we compared the opinions of skippers of large draggers (over 55 feet in length) with those of skippers of small draggers (55 feet and less in length) as illustrated in some of the previous tables. Surprisingly, there was no correlation between vessel size and opinions about the timing of the closure (Table 10), perceptions of stock conditions (Table 11) or preferred length of closure (Table 13). Large vessel skippers were no more likely than small operators to consider that the Gulf fishery should have been closed earlier or left open. Greater investments and the need to obtain more income to maintain their vessels appear unconnected to perceptions about the need for the closure or the length of the closure. Nor was there any difference in attitude towards the current condition of the stocks.

A notable difference between the larger and smaller draggers was found, however, when skippers were asked about the number of draggers that should be permitted in any future cod fishery. A much higher percentage of the large-scale operators (69.2 compared with 33.3) stated that there should be a reduction of the fleet to sixty-five draggers or less (Table 14). This reflects the common attitude among many skippers of large draggers that there should be fewer vessels so that vessels such as their own can more fully utilize their immense catching capacity. This also reflects their perception of state policy as favouring draggers as the future of the fishery—but a dragger fleet without many of the currently less viable participants. The fact that this proposed future closely resembles an undoing of the expansion of the dragger fleet in the early 1980s did not escape many of the skippers. Several skippers made comments like the following: "They [the DFO] caused this problem [the decline of the stock] when they let the new boats in [in the early 1980s]. Now they have to undo what they did or we're all sunk."

Table 14. Number of Draggers Desired in Reopened Fishery by Skippers' Vessel Size

	Number of Draggers (%)		
Vessel Size	*Up to 65*	*Over 65*	*Number*
Up to 55 feet	33.3	66.7	33
Over 55 feet	69.2	30.8	26
Total	49.2	50.8	59

Chi-square p. = .006, V= .36.

Regulation of the Future Fishery

Many dragger skippers not only saw the reversal of past mistakes as the key to the future fishery in terms of the number of draggers, but also in terms of the regulation of the fishery. All but one skipper accepted the need for quotas in the new fishery, but 70.5 percent would like to see a change in their administration. The majority of these changes involved a continuation of individual quotas, but with a considerable reduction in the transferability of these quotas. Individual quotas were supported by the vast majority of skippers because they did not want to return to the "rat race" that comes with open fleet quotas. Only a couple of skippers felt that enterprise allocations should be abolished because they reduced the incentive for extra effort: "The slightest bit of wind and no one leaves the dock." However, even though many skippers recognize advantages to transferability in circumstances where an owner is unable to fish and requires income, there was a pervasive negative attitude towards transferability in general.

The dominant reason for opposition to transferable licences was that they were felt to favour the skippers with more resources, even with the current rule that limits acquisition to the equivalent of double the starting quota for a 65-foot vessel. Owners of smaller vessels and people with fewer resources felt that transferable quotas threatened their own abilities to continue in the fishery, or at least that it would lead to excessive concentration of fishing capacity. Indeed, many small-boat dragger skippers felt that transferability was simply a means to force them out of the fishery. Several noted that any unused quota should go back to the government and that they "should just leave the fish in the water" or "let it swim away." Although such comments may show concern for conservation, the emphasis is almost always clearly on the perceived unfairness of allowing the larger or wealthier vessels more than their fair share. Hence, "leaving the fish in the water" is often seen as equivalent to spreading the unused quota among the entire fleet.

It was not only the skippers of smaller draggers who called for the abolishment of transferable licences. This view was also held by many skippers of larger draggers. For example, one of these skippers stated that even though he had personally benefitted from buying additional quota, he was against the continuation of the system on the grounds of fairness and social justice.

Dragger Activities during the Closure

Although the thoughts of many dragger skippers concerned their uncertain future, it would be inaccurate to think that the official cod moratorium meant that the draggers were tied up to the wharves and their crews were drawing compensation. The fleet actually continued to fish, and our survey found that all the skippers planned on fishing during 1994. In addition to not fishing for cod, the other crucial difference from past years was that the goal of the dragger fishers was to earn enough insurable weeks of income to qualify for unemployment insurance benefits (a number that was raised from ten weeks to twelve weeks in 1994).

The two key species for draggers were shrimp and scallop, with some boats, especially those with mid-water trawls, also hoping to catch redfish. The vast majority of the 68.9 percent who held shrimp licences planned to rely heavily on this resource. Because of the increased significance of shrimp, a new limit on each vessel's yearly catch was imposed to help spread out this resource more than in the past. The opening date of the shrimp fishery was delayed, however, for over one week because of the extremely low prices paid for the first catches. The low prices were claimed by the buyers to be in response to the small size of the shrimp that were landed. Dragger skippers, however, disagreed considerably among themselves about the actual size of the shrimp. Although some of this disagreement might have been the result of actual differences in the size of shrimp caught by different boats, or the use of a new grading system, one dragger skipper saw a clear pattern related to his perception that the state planned to reduce the size of the fleet. He suggested that the disagreements over the size of the shrimp correlated with the division between dragger skippers living in Port au Choix and those in other areas: "The Port au Choix boats are mainly paid off, see. Therefore, they want to force the other shrimpers out of business by messing up the shrimp fishery. That's why they are the one's saying the shrimp is small." Whether or not this suggestion is true, it reflects the growing opinion that only some of the draggers are to be part of the state's fishery of the future. Despite these problems at the beginning of the fishery, the shrimp quota was caught by early summer at a respectable average price of fifty-five cents per pound (*Northern Pen,* August 2, 1994). Even though this made for a successful season, most of the shrimp draggers were only able to attain around seven weeks of insurable income from the fishery.

Those vessels that lacked a shrimp licence but were among the 42.6 percent of the fleet that held a scallop licence concentrated on that fishery. This brought them into competition with the longliner fishers who had been pursuing this species for the past decade. The price of scallops had again risen to over seven dollars per pound by the end of 1993 and had drawn nearly seventy vessels back into the fishery. This meant that the new otter-trawl entrants were adding to a fishery that was already under heavy harvesting pressure. They were also now competing against many of the longliner fishers who had been previously chased from the cod fishery by the draggers. This resentment towards the draggers remained subdued, however, and many longliner scallop fishers expressed the sentiment of one skipper who said, "Well, I don't like it [having the draggers in the scallop fishery], but a man has to do whatever he can. If I were in their place, I'd do the same thing." As a result of the increased competition, most of the scallop boats earned about seven weeks of income at an average price of $7.25 per pound. Hence, many of the dragger skippers had also to pursue redfish to earn the remaining number of weeks of income to qualify for UI.

Although some boats managed to supplement their 1994 income with redfish, this species began to face quota cuts of its own. The announcement of a reduction of approximately 75 percent to the redfish quota for 1995 led to an increase in dragger skippers attempting to sell their boats. Some skippers of larger vessels with mid-water trawls considered switching to smaller, less expensive draggers, while some older skippers simply wanted out of the fishery (*Northern Pen,* November 8, 1994). When a limited buyout program was announced by the government in 1995, it was greeted unenthusiastically by many skippers, and most remained uncertain about their future in the fishery.

Conclusion

Although the Gulf moratorium should not be seen as a policy move designed exclusively to deal with the dragger fleet, local residents of all occupations realized that the actions taken by the state during the closure would generate its clearest stand on the issue of the dragger fleet and the future of the fishery. The moratorium was obviously an attempt to save *the fishery,* but most fixed-gear fishers saw few if any signs during the first few months of the closure that the state was interested in saving *their fishery.* However, most dragger skippers were convinced that the state would attempt to reduce the size of the fleet during the closure. The question was how, and by how much. This question would be answered by policies such as those on future buybacks, leniency on loan payments and the regulation of alternative fisheries. The state's reluctance to take any quick, dramatic steps at the beginning of the moratorium suggested that the state was still attempting to reduce the dragger fleet while taking as little responsibility for this reduction as possible.

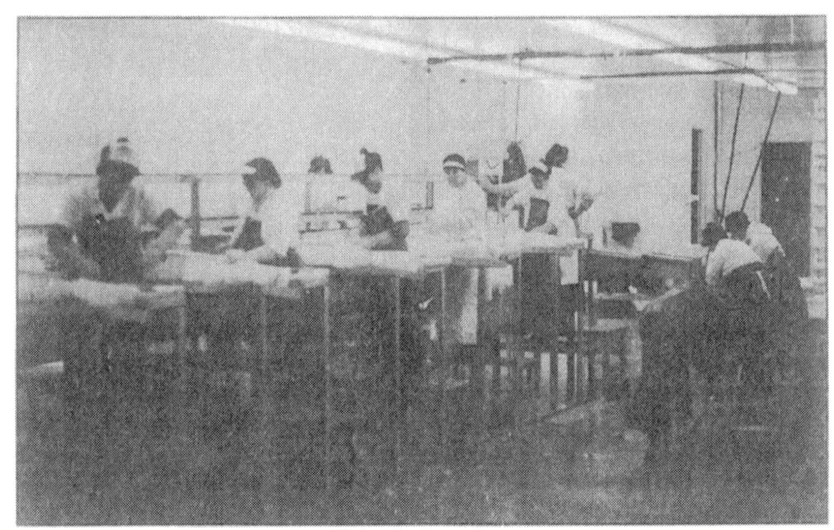

Fish plant workers

Boy cutting cod tongues

7
The Future of Domestic Commodity Production

The cod moratorium represents a convenient point at which to evaluate the fate of domestic commodity production in northwest Newfoundland by taking into account what the various influences over the previous thirty years had produced. Are domestic commodity producers being absorbed into and transformed by a capitalist system? Have the dragger skippers become petty capitalists? If they have, are there signs that these petty capitalists will either expand into, or be absorbed into, a larger-scale capitalist fishery? Our conclusion is that DCP is in a precarious position and is likely to become a minor form of the social organization of fishing in the future.

The Current Fishery: Participation Rates

Table 15 shows an expansion in the number of vessels along the northwest coast of Newfoundland (which corresponds to fishing areas M and N on Map 1) from 1984 to 1993, that is, from the most successful time to the point of collapse. This finding appears inconsistent with the general decline in the fishery we have documented. An examination of the size of vessels involved in this increase, however, appears to resolve this paradox:

> Distribution of the increase by vessel size in 4R, 3Pn reflects the declining economic prospect of those involved in the groundfish fishery. While there has been . . . a[n] increase in vessels less than 35' . . . [there has been a] decline in vessels greater than 35'. . . . Present policy permits full-time fishermen to register additional vessels less

Table 15. Fisheries Participation along the Northwest Gulf Coast (Cape Gregory to Cape Norman), 1984–93

	1984	*1993*	*% Change*
No. of full-time fishers	1408	1463	3.9
No. of part-time fishers	1652	1911	15.7
Total fishers	3060	3374	10.3
Vessels <35 feet	1362	1464	7.5
Vessels 35–65 feet	152	136	-10.5
Total vessels	1514	1600	5.7

Source: DFO, St. John's.

> than 34'11". Since no increase in fishing effort has occurred, it is felt the registration of auxiliary vessels accounts for the increase (Marshall 1990:29).

For the region of prime concern to us, small boats increased by 7.5 percent, while larger vessels fell by 10.5 percent (Table 15). As Marshall suggests, the decline in vessels over 35 feet in length reflects the unwillingness of fishers to invest heavily in a declining resource, although additional data (supplied by the DFO, St. John's) show an increase in this size class in the core ports of Port au Choix and Port Saunders. That is, this pessimistic attitude seems to be unevenly distributed along the coast. We further suggest that the increase in the numbers of small boats and fishers, especially those active part-time, indicates a continued perception of the small-boat fishery as a last resort "safety net" response to the dismal state of the fishery and the economy in general. With no incentive to invest in a declining fishery and no alternative sources of income, the small-boat fishery appears still to be perceived as one of the few constant sources of income, i.e., as part of a rational, adaptive strategy. Because of the relatively small amounts of capital involved, and the fact many residents possess at least the minimum skills required, small-boat fishing is a potential source of income for many families in the area. This perception is far from an outdated mythology held over from earlier times. Given current government policies, a strategy of fishing for the few weeks necessary to qualify for a "make-work" program that in turn allows one to qualify for UIC is a plausible, although far from desirable, economic plan. At least, it is rational to acquire the necessary boat and licence in order to have such an insurance policy. Perhaps this is why, as one provincial fishery officer stated, on the Great Northern Peninsula, "everybody and their dog has a small boat!"

In any case, the statistics on numbers of fishers and vessels we have presented in this book fail to show either the small-boat fishery based on DCP or the petty-capitalist dragger fleet headed towards imminent extinction when the 1994 moratorium on cod fishing was instituted. This might imply that both the DCP-based fixed-gear fishery and the petty-capitalist dragger fishery could be sustainable, given the return of the cod fishery. At first glance this analysis of northwest Newfoundland supports Sinclair's 1985 conclusion that DCP can continue to exist, even while some entrepreneurial fishers create a petty-capitalist form of production. Such a conclusion, however, must be tempered by several observations. First, the continued existence of both sectors of the fishery at the time of the cod moratorium may simply be the result of the state, in an attempt to minimize social protest, having continued to postpone certain inevitable decisions, such as drastic changes in the UIC system, that would otherwise already have spelled the end for DCP in the area. Indeed, despite two years of a moderate cod-trap fishery, the fixed-gear fishery still appeared to be far from "sustainable," and trends in landings were pointing towards the demise of the small-boat, fixed-gear fishery. Second, hidden beneath the caricature of fixed-gear fishers as domestic commodity producers and draggers as petty capitalists are subtleties in social organization, especially within the crew structure of the dragger fleet. Third, the reality is that the cod moratorium did occur and the future of the draggers as well as the fixed-gear fishers is in jeopardy. This raises the question of exactly what was the relationship between these two types of production, the role of the state towards them and the collapse of the Gulf cod stocks.

Are Draggers Really Petty Capitalist?

Of particular theoretical importance in evaluating the fate of DCP in northwest Newfoundland is the extent to which the draggers really represent petty-capitalist enterprises instead of domestic commodity production. The two key variables in determining the classification of dragger fishers as petty capitalists are their reliance on kin as labour and their commitment to an expanded, as compared to a simple, form of production.

In contrast to the "immense pressure to find a place for each adult male in the close kin network" that dominated the crew structure of DCP, Sinclair reported that by the early 1980s the dragger skippers of Port au Choix insisted that "kinship is irrelevant: 'You choose the best man for the job'" (1985:97). Although this would appear to be a clear indication of a shift from DCP to petty capitalism, Sinclair found that:

> Nevertheless, the "best man" frequently turns out to be a son, brother or cousin. Of 57 crew members on Port au Choix draggers, 22 were agnatic kinsmen of the skipper and another nine were related by marriage. Only five vessels were crewed entirely by non-kinsmen (1985:97).

Given the ambiguous nature of preference for kin in the dragger fleet of the early 1980s, Sinclair (1985:97) hypothesized that the key would be whether or not kin continued to dominate crew structures in the future, especially when sons, who may not be the "best men," come of age.

An analysis of dragger crew structures in 1994 revealed continued reliance on kin as crew members. Table 16 indicates that skippers employed a wide variety of kin in their crews, with 60 percent having at least one son on board. Further, Table 17 indicates that well over half of the fleet (61 percent) were crewed exclusively by kin. Table 17 also shows that this dependence on kin was not evenly distributed throughout the fleet. Smaller vessels (55 feet and under) were much more reliant on kin than were the larger vessels (56–65 feet). Indeed, over three-quarters (75.8 percent) of the smaller vessels were crewed exclusively by kin. There was also an indication that the skippers of smaller vessels hire more kin than the vessels really require. Perhaps this pattern of hiring kin was influenced by the desire to qualify as many kin as possible for UI. Whereas 36 percent of larger vessels had crews of five or more, 48.5 percent of small-boat crews fell into this category. This finding is surprising, as we would have expected smaller boats to have smaller crews. This evidence of kin-based crews suggests that there is considerable diversity in the extent to which dragger enterprises have shifted from a DCP-type of social organization to one more characteristic of capitalism.

The presence or absence of kin in crews is just one part of the relationship between kinship and the transition to a more capitalistic form of production. There is also the question of the relationship between crew members and how this is influenced by kinship. Much of the variation in this dimension of the capitalist nature of the dragger fleet correlates with the geographic division between the Port au Choix area and the Straits (i.e., the Strait of Belle Isle). Although the share system of payment has persisted throughout the fleet, most of the crews in the Port au Choix area quickly abandoned the co-ownership and egalitarian authority patterns among brothers that were common in the tradi-

Table 16. Percentage of Skippers with Number of Specified Kin in Crew (N=60)

	Number in the Crew (%)			
Relationship	*None*	*One*	*Two*	*Three or More*
Brothers	43.3	18.3	20.0	18.4
Sons	40.0	35.0	15.0	10.0
Father	93.3	6.7	NA	NA
Brother-in-law	80.3	18.0	0.0	0.0
Son-in-law	93.3	5.0	1.7	0.0
Nephew	85.0	10.0	5.0	0.0
Uncle	98.4	1.6	0.0	0.0
Cousin	88.3	10.0	1.7	0.0
Other relatives	85.0	5.0	5.0	5.0
Non-kin	61.7	21.7	11.7	5.0

Table 17. Presence of Crew Unrelated to Skipper by Vessel Size

	Presence of Unrelated Crew (%)		
Vessel Size	*None*	*One or More*	*Number*
Up to 55 feet	75.8	24.2	33
Over 55 feet	42.3	57.7	26
Total	61.0	39.0	59

Chi-square p. = .009, V= .34.

tional DCP form of production (see Firestone 1967) and still characterize the small-boat fishery. Two reasons for this switch were the huge increases in investment and the greater differentiation between the work roles of skippers and deckhands. The large monetary investment meant that an economic responsibility was placed on the owner of a vessel, and this was thought to require an accompanying amount of decision-making power. The very nature of the dragger technology also entailed a more strict and differentiated division of labour. The duties of steering the boat, often through congested areas and/or ice, and interpreting the subtle radio conversations that might indicate fish in certain areas, meant that it was usually impractical for skippers to "be at the fish" even if they desired to be. This also meant that the duties of deckhands and skippers, which were often indistinguishable in small-boat fisheries, became radically different. As one crew member stated during the winter fishery, "Out here [on deck] it's just a matter of brute strength; in there [the wheelhouse] it's all mind games."

The stress that these structural changes place on traditional kinship-based patterns of organization is perhaps even clearer when one examines the attempts by dragger crews in the Straits to maintain traditional kinship patterns of authority. For example, skippers from Port au Choix often express their amazement at the attempt by crews from the Straits to maintain co-ownership among a set of two to five brothers. As one skipper stated, "I don't know how they do it; they must be crazy." The main source of amazement concerns decision-making. Even though one brother is always designated the skipper, it is thought to be unreasonable for another brother with an equal amount of money invested in the operation always to defer.

Discussions with dragger fishers from the Straits confirm that it is difficult to apply traditional work organization patterns to the dragger fleet. One fisher described the role of being the skipper of his crew of co-owning brothers as simply "hell." Often these difficulties lead to complex and contradictory patterns of authority on vessels:

> Well, we are all owners, because we all put up the money the same and take the same shares, but [one fisher] is the "skipper," because he steers the boat, but [another fisher] makes most of the decisions on deck because he has the most experience, but [a third fisher] is the "boss" and makes the money decisions because he's the oldest.

There are also examples of incompatibility between the dragger technology and DCP forms of social organization so great that crews are forced to split apart: "All of us [brothers] used to fish together, but we each wanted to do things our own way, so we split apart." Even in these situations, however, brothers are expected to have an informal special right to rejoin a crew even if it is not economically advantageous for the crew. Comments like "I hope my brother can get on with someone else, because we already have a full crew" were common.

When the criterion of kin labour is used to classify the dragger fleet as DCP or petty capitalist, an ambiguous answer is obtained. The very nature of the technology clearly encourages a capitalist form of organization and production, and this has certainly pushed many crews far towards the petty-capitalist position. There is also, however, considerable evidence of DCP forms of relationship tenaciously enduring despite their incompatibility with the technology. Indeed, some dragger crews have retained many of the characteristics of DCP.

The view that there are major differences in the extent to which individual draggers have become petty-capitalist operations is also supported when one examines the question of simple versus expanded reproduction. Although the history of the fleet as a whole unquestionably shows a pattern of expanded reproduction, there is considerable variation among skippers. On the one hand, a number of dragger skippers are clearly capitalists committed to expanded reproduction. For example, in 1991, a skipper of a 55-foot dragger stated the following plan for future expansion: "I'm thirty-five [years old] now and skipper of my own boat. That's pretty good, but it's just a 55-footer. I hope to have a 65-foot steel [boat] by the time I'm forty, and maybe get a mid-water trawl on it before I'm forty-five." Plans for future expansion are not limited simply to larger vessels. Another skipper stated an even more ambitious ultimate goal: "I became a skipper so I wouldn't have to wear rubber clothes [used by crew members on the deck], but my real goal is to own a big dragger [over 65 feet] so I'll never have to go to sea." There are also a few skippers who have used the capital gained from dragging to invest in property and businesses both related and unrelated to the fishing industry. In these cases, the term *capitalist* surely applies, and they might even indicate some expansion from petty capitalism to larger-scale capitalistic forms of production.

The term *expanded mode of reproduction* fails, however, to fit the economic strategies of many of the smaller draggers. Conversations among these crews and their families bear striking resemblances to those of small-boat crews who are clearly still engaged in simple reproduction. The dragger fishers may be talking about subjects, such as large bank loans and insurance payments, that are unfamiliar to small-boat fishers, but the goal of the discussions in both cases is simply to make it through one more year with what one already has. Dreams of expanded reproduction were almost certainly present during the transition to dragger technology, but for many dragger crews the goal of expanded reproduction has been replaced with the goal of simple reproduction so familiar to their ancestors.

Because of the factors just discussed, it is inaccurate to equate the fate of the dragger fleet as a whole with the fate of petty capitalism in northwest Newfoundland. It is more accurate to separate the dragger fleet into those operations that are clearly capitalist in organization and those that still approximate the criteria of DCP, and then to examine the fate of each category separately. Instead of indicating more of a future for DCP in northwest Newfoundland because some of the dragger fleet may fit the criteria of DCP, this reclassification actually suggests a clearer pro-petty capitalist and anti-DCP position on the part of the state than does an analysis equating all draggers with petty capitalism. This is because many of the state's seemingly ambivalent and contradictory policies over the previous two decades coincide with the view that it favoured the section of the dragger fleet that was composed of more clearly capitalist enterprises. Perhaps this is what was sensed by the small-boat dragger skipper who stated that the real division is between the big draggers and everyone else.

If one interprets the past decade of state policy as favouring the most capitalistic members of the dragger fleet over both the DCP draggers and small-boat fishers, the question still remains as to whether the petty-capitalistic draggers were on the road to being absorbed or expanded into larger-scale capitalist enterprises. Although there had been significant expansion in the capital intensity of individual vessels over the period, this appeared to have reached its limit because of size restrictions on vessels. It is not inconceivable, however, to speculate that certain changes, perhaps relaxing the limits in the amount of quota that could be accumulated under the ITQ system, might have produced larger-scale capitalist enterprises. This could have taken the form of either multiple-vessel ownership or vertical integration with other organizations in the industry. The point of such hypothetical speculation is that, should the cod stocks recover, such possibilities remain options for the future.

DCP, Petty Capitalism, the State and the Collapse of the Cod Stocks

Cold water temperatures, increased numbers of seals and other ecological factors may have contributed to the collapse of the Gulf cod stock. Overfishing, including overfishing by the northwest Newfoundland dragger fleet, surely contributed most to this collapse, although we may never know the exact balance of forces for sure. Further, the "blame" for this overfishing is a complex issue that we have no desire to pursue here. A more productive question concerns the role the diminishing Gulf cod stock played in the fate of DCP and petty-capitalist fishing enterprises. Perhaps this role can be seen most clearly by hypothesizing what would have happened had the Gulf cod stock not started to diminish in the mid-1980s. Although the exact effects can never be known, it seems reasonable to assume that the successful dragger fleet might have continued to expand as a result of increased pressure for more licences and intensified capitalization of existing enterprises. Given the impact the dragger fleet was apparently having on the fixed-gear catches of the mid-1980s, such an expansion of the dragger fleet would probably have been accompanied by a more pronounced trend towards the demise of the fixed-gear fishery. Such a demise might have been accelerated by the state's more wholehearted support for the successful dragger fishery as the sole fishery of the future. Hence, the collapse of the Gulf cod stock might actually have contributed to the continued existence of DCP in northwest Newfoundland. Again, the point of such speculation is that the future of the fishery is still to be determined and may depend on the re-evaluation of some previous goals of the state that never materialized because of the collapse of the cod stocks.

Consistent with the hypothesis that the collapse of the cod stock prevented the materialization of a clearer trend towards increased capitalization and the demise of DCP in northwest Newfoundland is the role of the state's management policies during the last decade. By far the most fundamental policies were the enterprise-allocation and ITQ regulations for the dragger fleet. Despite the various factors influencing the original adoption of enterprise allocations, the previous analysis of the years following their introduction makes it clear that they quickly became a tool to "rationalize" the dragger fishery into a smaller but more capitalistic industry. Hence, there is little to suggest that the fishery envisioned for the future included much of a role for DCP, regardless of the type of technology they employed.

The enterprise-allocation and ITQ systems of management failed to produce an economically viable fishery and a sustainable resource base for a rationalized capitalist fishery for two main reasons. First, the inadequacy of these management policies lay less with their theoretical assumptions than with the failure to enforce them. Not only did illegal fishing practices play a role in depleting the cod stocks, but they also diminished the rationalizing effect of the policies because inefficient enterprises were able to remain in the fishery through illegal tactics. This brings up the second reason for the inability of enterprise allocations and ITQs to produce a sustainable and rational fishery: the rationalizing force of the policies was never allowed to take full effect. The restrictions on the original enterprise allocations meant that they were more likely to keep vessels in the fishery instead of weeding them out. Even when the enterprise allocations became ITQs they were still subject to strict limitations on quota restrictions. User fees for monitoring expenses were also only gradually and moderately introduced. The important point is that the ITQ system, at least after 1987, was clearly *designed* to rationalize the fishery. The restrictions on its ability to do so appear to be simply appeasements made to residents still dependent on DCP in order to minimize the type of social protest that could jeopardize the state's policy.

There are other indications that there will be little room for DCP in the future of northwest Newfoundland. One such indication is the support, from both the state and the FFAW, for the "professionalization" of the fishery. The supporters of professionalization stress things such as greater education and training, more concern for safety and quality, and a general increase in the social status of "fishers." Often neglected in these discussions is the fact that a crucial aspect of professionalization is the exclusion from the fishery of many of its current members. Although this is sometimes made palatable by referring to the excluded fishers as "moonlighters," some residents sense the impact that "professionalization" would have on the traditional way of life:

Resident: I don't know what's on the mind of government right now, but I certainly don't agree with what they say about getting rid of the fishermen, the inshore [small-boat] fishermen.

Interviewer: You mean the "professionalization?"

Resident: Yes, boy, we can't live that way. We just got to leave. If the people [small-boat fishers] haven't got an option here to sometimes go and work for a contractor and make a few bucks . . . if that option is gone, then he's got no choice, only to leave. He got to leave. You can't have in the Strait of Belle Isle that system [of only "professional" fishers]. You just can't have it.

The retraining program, which has made up a major part of the government's policy for dealing with the moratorium, is also clearly an attempt to ease many residents out of the fishery and out of DCP. However, the clearest single determinant of the future of DCP in northwest Newfoundland will not have to do directly with fishing regulations. Instead, the fate of DCP may hinge on the changes the state makes in the UIC system. Because UI was probably the chief reason DCP continued to exist in northwest Newfoundland over the past thirty years. As mentioned in Chapter 4, there has been a growing belief that the current UIC system will be modified because it is simply too expensive (although it is rare, in fact, that expenditures from the UI fund are not covered by contributions when we consider Canada as a whole). The extent and form of these revisions will probably tell more about the future of the fishery than any other state regulation affecting the fishery in the near future. The importance of the UIC system for the maintenance of DCP, and the realization that this system might come to an end in the near future, was illustrated in the comments of one resident. Gazing at several hundred expensive snowmobiles gathered on one of the frozen harbours for the "winter carnival" races, she stated, "If UI stops, it won't be long 'til people start looking for ways to eat those machines." Drastic cutbacks in the UIC program may be the simplest way for the state to rationalize the fishery, and the transitional context of the moratorium may be the most strategic opportunity for the state to take such action.

Conclusion

We suggest that the continued existence of DCP in northwest Newfoundland is misleading both in a theoretical and practical sense. From the perspective of theoretical arguments about the fate of DCP in capitalist societies, it is misleading because many of the historical trends point towards the demise of DCP in northwest Newfoundland. The continued existence of DCP may only be the result of (1) the state trying to make the disappearance of DCP as gradual as possible to avoid social protest; (2) the collapse of the resource base upon which the transitional capitalist enterprises could have continued to expand; and (3) simply the failure to enforce policies designed to eliminate DCP.

From the practical perspective of individuals living in the area, the conclusion that DCP is a *sustainable* way of life in northwest Newfoundland is even more misleading. Unless major changes in policy goals occur during the current moratorium, there is every reason to expect that there will be little room for DCP in the fishery of the future. One of the clearest signs that this view is shared by residents is evident in the plans of young people. Among the forty-three graduating seniors from a local high school in 1991, only two young men said their future plans were "to fish." Both of them were sons of dragger skippers. If this is indeed the future of the fishery, it may be regrettable in the sense that it is not what many of the residents would have chosen. As one graduating senior stated, "I'd prefer to live here, but I sure hope to be gone within five years." However, the traditional DCP lifestyle of northwest Newfoundland should not be idealized. It was often a way of life that meant severe economic hardship. As one dragger skipper, who had participated in all of the transitions that had taken place since 1960, stated, "Sure we could go back to a small-boat fishery, but I for one don't intend to have to starve again."

Appendix: Data Collection Methodology

Sinclair's research in Port au Choix in 1981 and 1982 consisted of participant observation and interviews with skippers, crew members and plant officials, supplemented by historical documents and secondary sources, including a DFO survey of the fleet that provided detailed information on crew structure and economic performance. Statistical information on fisheries and social characteristics were collected for the entire peninsula and almost every harbour was visited. His work focused on the mobile-gear cod and shrimp fisheries. In 1988, Sinclair and Lawrence F. Felt conducted a survey of 250 households in which 554 persons were interviewed on a wide variety of issues concerned with how people coped with living in an isolated, economically marginal area. Local persons were hired to conduct all interviews.

Palmer's fieldwork from May 1990 to May 1992 consisted of over 200 formal interviews with residents of northwest Newfoundland and southern Labrador, and over 20 formal interviews with federal and provincial fisheries officers, fish plant operators, FFAW representatives and various other professionals associated with the fishery. Palmer chose his village of residence because it adjoined the one Firestone (1967) had studied twenty-five years earlier. The interviewing of residents included an unstructured survey during the summer of 1990 and a structured survey of 171 adult local residents of one Great Northern Peninsula community during the fall of 1990 (see Chapter 4). In addition to interviewing, Palmer also engaged in extensive participant observation in both the fixed-gear and mobile-gear fisheries. This included accompanying crews from Port au Choix to Eddie's Cove East as they pursued capelin, fixed-gear cod (cod-trap and gill-net), mobile-gear cod, mobile-gear shrimp, mobile-gear red fish, fixed-gear turbot, scallop and lobster.

The 1994 survey of otter-trawl skippers was designed by Palmer and Sinclair and administered by mail, telephone or in-person interviews by Palmer. It produced data concerning the cod moratorium on sixty-one of the seventy-six current otter-trawl skippers.

Lobster fishing camp—St. John Bay

References

Andersen, Terry L., and Peter J. Hill. 1991. "Privatizing the commons: an improvement?" *Southern Journal of Economics* 50:438–50.

Apostle, Richard, and Gene Barrett with others. 1992. *Emptying Their Nets: Small Scale Capital and Rural Industrialization in the Nova Scotia Fishing Industry.* Toronto: University of Toronto Press.

Bailey, Conner, John Bliss, Glenn Howze, and Larry Teeter. 1993. "Dependency theory and timber dependency." Paper presented to Rural Sociological Society, Orlando, Florida, August.

Bailey, Conner, Peter Sinclair, John Bliss, and Karni Perez. 1996. "Segmented labor markets in Alabama's pulp and paper industry." *Rural Sociology* 61:475–496.

Binkley, M.E. 1989. "Nova Scotian offshore groundfish fishery: effects of the enterprise allocation and the drive for quality." *Marine Policy* 13(4):274–84.

Black, William A. 1966. *Fishery Utilization, St. Barbe Coast, Newfoundland.* ARDA study no. 1043. Ottawa: Department of Mines and Technical Surveys.

Bonnano, A., L. Busch, W. Friedland, L. Gouveis, and E. Mingioni, eds. 1994. *From Columbus to ConAgra: The Globalization of Agriculture and Food.* Lawrence: University of Kansas Press.

Boyd, Rick O., and Christopher M. Dewees. 1992. "Putting theory into practice: individual transferable quotas in New Zealand's fisheries." *Society and Natural Resources* 5:179–98.

Brothers, Gerald. 1975. *Inshore Fishing Gear and Technology.* St. John's: Environment Canada, Fisheries and Marine Service.

Buttel, Frederick H., and Pierre LaRamee. 1991. "The 'disappearing middle': a sociological perspective." In William H. Friedland, Lawrence Busch, Frederick H. Buttel and Alan P. Rudy (eds.), *Towards a New Political Economy of Agriculture.* Boulder, Colo.: Westview.

Campbell, Capt. Charles. 1888. "The Lobster fisheries on the west coast." In D.W. Prowse, *A History of Newfoundland from the English, Colonial and Foreign Records.* London: Macmillan (1895).

Canada. 1983. *Enterprise Allocations for Inshore Mobile Gear Groundfish Fleet 19.8 m or Less in NAFO Divisions 4R, 3Pn.* Corner Brook, Nfld.: Department of Fisheries and Oceans.

—. 1990. *Report of the Working Group on the 4R, 3Pn Cod Fishery.* Moncton, N.B.: Department of Fisheries and Oceans, Gulf Region.

Canadian Broadcasting Corporation (CBC) Radio. "Fisheries broadcast." February 4 and 8, 1991; February 19, 1992:April 13, 1992.

Canadian Broadcasting Corporation (CBC). 1990. "The Fishery of the '90s." Television special. October 22.

Canning and Pitt Associates. 1992. "Assessment of the Fisheries Downturn on the Great Northern Peninsula." Prepared for the Great Northern Peninsula Central and Nortip Community Futures Committees.

Cashin, Richard (chair). 1993. *Charting a New Course: Towards the Fishery of the Future.* Task Force on Incomes and Adjustment in the Atlantic Fishery. Ottawa: Supply and Services Canada.

Clement, Wallace. 1986. *The Struggle to Organize: Resistance in Canada's Fishery.*

Toronto: McClelland and Stewart.

Copes, Parzival. 1986. "A critical review of the individual quota as a device in fisheries management." *Land Economics* 62(3):278–91.

Crothers, S. 1988. "Individual Transferable quotas: The New Zealand experience." *Fisheries* 13(1):10–12.

Crowley, R.W., and H. Palsson. 1992. "Rights Based Fisheries Management in Canada." *Marine Resource Economics* 7(2):1–21.

Davis, Anthony. 1984. "Property Rights and Access Management in the Small Boat Fishery: A Case Study from Southwest Nova Scotia." In C. Lamson and A.J. Hansen (eds.), *Atlantic Fisheries and Coastal Communities.* Halifax: Dalhousie Ocean Studies Program.

—. 1991. *Dire Straits. The Dilemmas of a Fishery: The Case of Digby Neck and the Islands.* St. John's: Institute of Social and Economic Research.

Dewar, Margaret E. 1990. "Federal intervention in troubled waters: lessons from the New England fisheries." *Policy Studies Review* 9 (Spring):485–504.

Dewees, Christopher M. 1989. "Assessment of the implementation of individual transferable quotas in New Zealand's inshore fishery." *North American Journal of Fisheries Management* 9(2):131–39.

Eder, James F. 1993. "Family farming and household enterprise in a Philippine community, 1971–1988: persistence or proletarianization?' *Journal of Asian Studies* 52:647–71.

Eythorsson, Einar. 1996. "Coastal Communities and ITQ management: The case of Icelandic fisheries." *Sociologia Ruralis* 36:212–23.

Felt, Lawrence F., and Peter R. Sinclair, eds. 1995. *Living on the Edge: the Great Northern Peninsula of Newfoundland.* St. John's: Institute for Social and Economic Research.

Finlayson, Alan Christopher. 1994. *Fishing for Truth: A Sociological Analysis of Northern Cod Stock Assessments for 1977–1990.* St. John's: Institute for Social and Economic Research.

Firestone, Melvin M. 1967. *Brothers and Rivals: Patrilocality in Savage Cove.* St. John's: Institute of Social and Economic Research.

Fisher, C.F. 1977. *An Economic Assessment of the Newfoundland Shrimp Fishery, 1976.* St. John's: Department of Fisheries and Environment.

Fisher, C.F., B.P. Ferguson, and M. Vaillancourt. 1980. *Task Force Report on the Economic Performance of Fleets Engaged in the Gulf of St. Lawrence Shrimp Fisheries.* Ottawa: Department of Fisheries and Oceans.

Found, H.R. 1963. *Production and Processing of Cod in Rural Communities of Newfoundland.* St. John's: Department of Fisheries.

Frechet, Alain. 1991. "Evaluation du Stock du morue du nord du Golfe de Saint Laurent (divisions de L'OPANO 3Pn, 4R et 4s) en 1990." CAFSAC research document 91/43.

Friedland, William H. 1991. "Introduction: shaping the new political economy of advanced capitalist agriculture." In William H. Friedland, Lawrence Busch, Frederick H. Buttel and Alan P. Rudy (eds.), *Towards a New Political Economy of Agriculture.* Boulder, Colo.: Westview.

Friedland, William H., Lawrence Busch, Frederick H. Buttel, and Alan P. Rudy, eds. 1991. *Towards a New Political Economy of Agriculture.* Boulder, Colo.: Westview.

Friedmann, Harriet. 1980. "Household production and the national economy: concepts

for the analysis of agrarian formations." *Journal of Peasant Studies* 7:158–84.
Gordon, H. Scott. 1954. "The Economic Theory of a Common Property Resource: The Fishery." *Journal of Political Economy* 62:124-42.
Haedrich, Richard L., and Johanne Fischer. Forthcoming. "Stability and change of exploited fish communities in a cold ocean continental shelf ecosystem." *Senckenbergiana maritima*.
Hanrahan, Maura. 1988. *Living on the Dead: Fishermen's Licensing and Unemployment Insurance Programs in Newfoundland.* St. John's: Institute of Social and Economic Research, research and policy paper no. 8. St. John's: ISER.
Hardin, Garrett. 1968. "The Tragedy of the Commons." *Science* 162:1243–48.
Hare, F. Kenneth. 1952. "The climate of the island of Newfoundland: a geographical analysis." *Geographical Bulletin* 2:36–88.
Hay, David A. 1992. "Rural Canada in transition: trends and developments." In David A. Hay and Gurcharn S. Basran (eds.), *Rural Sociology in Canada.* Don Mills: Oxford University Press.
House, J. Douglas with S.M. White and P. Ripley. 1989. *Going Away . . . and Coming Back: Economic Life and Migration in Small Canadian Communities*. Institute of Social and Economic Research, report no. 2. St. John's: ISER.
Hutchings, Jeffrey A., and Ransom A. Myers. 1994. "What can be learned from the collapse of a renewable resource? Atlantic cod, Gadus morhua, of Newfoundland and Labrador." *Canadian Journal of Fisheries and Aquatic Sciences* 51:2126–46.
Inglis, Gordon. 1985. *More than Just a Union: The Story of the NFFAWU.* St. John's: Jesperson.
Jentoft, Svein. 1989. "Fisheries co–management: delegating government responsibility to fishermen's organizations." *Marine Policy* 13:137-54.
—. 1993. *Dangling Lines: The Fisheries Crisis and the Future of Coastal Communities: The Norwegian Experience.* St. John's: Institute of Social and Economic Research.
Jentoft, Svein, and Trond I. Kristoffersen. 1989. "Fishermen's co-management: the case of the Lofoten fishery." *Human Organization* 48:355–86.
Kennedy, John C. 1996. *Labrador Village.* Prospect Heights, Ill.: Waveland.
Larson, Bruce A., and Daniel W. Bromley. 1990. "Property rights, externalities, and resource degradation: locating the tragedy." *Journal of Development Economics* 33:235–62.
LeDrew, Bevin R. 1988. *A Study of the Conflict between Fixed and Mobile Gear in Western Newfoundland.* St. John's: Department of Fisheries and Oceans.
Libecap, Gary. 1989. *Contracting for Property Rights.* New York: Cambridge University Press.
Lind, Christopher. 1995. *Something's Wrong Somewhere: Globalization, Community and the Moral Economy of the Farm Crisis.* Halifax: Fernwood.
Llambi, Luis. 1988. "Small modern farmers: neither peasants nor fully-fledged capitalists?' *Journal of Peasant Studies* 15:350–72.
Long, Norman. 1984. "Creating Space for Change: A Perspective on the Sociology of Development." *Sociologia Ruralis* 24:168–73.
Maclean's. 1991. "Fisheries Riot." 104(12):15.
Marchak, M. Patricia. 1983. *Green Gold: The Forest Industry of British Columbia.* Vancouver: University of British Columbia Press.
—. 1991. "For whom the tree falls: restructuring of the global forest industry." *BC Studies* 90:3–24.

Marshall, J.M. 1990. "Report on the Working Group on the 4R, 3Pn Cod Fishery." Moncton, N.B.: Department of Fisheries and Oceans, Gulf Region.

McCay, Bonnie J., and James M. Acheson, eds. 1987. *The Question of the Commons: The Culture and Ecology of Communal Resources.* Tucson: University of Arizona Press.

McCay, B.J., and Carolyn Creed. 1990. "Social Structure and Debates on Fisheries Management in the Mid-Atlantic Surf Clam Fishery." *Ocean and Shoreline Management* 13:199–229.

McGoodwin, James R. 1990. *Crisis in the World's Fisheries: People, Problems, and Policies.* Palo Alto, Calif.: Stanford University Press.

Neary, Peter. 1980. "The French and American shore questions as a factor in Newfoundland History." In J. Hiller and P. Neary (eds.), *Newfoundland in the Nineteenth and Twentieth Centuries: Essays in Interpretation.* Toronto: University of Toronto Press.

Newfoundland. 1857. *Census of Newfoundland.* St. John's.

Newfoundland Fisheries Board. *1940–48. Annual Reports.* St. John's: Newfoundland Fisheries Board.

Northern Pen. February 11, 1982; July 27, 1990; September 19 and 26, 1990; July 19, 1991; August 2, 1994; November 8, 1994.

Omohundro, John. 1994. *Rough Food: The Seasons of Subsistence in Northern Newfoundland.* St. John's: Institute of Social and Economic Research.

Ostrom, Elinor. 1990. *Governing the Commons: The Evolution of Institutions for Collective Action.* New York: Cambridge University Press.

—. 1992. "Community and the endogenous solution of commons problems." *Journal of Theoretical Politics* 4(3):343–52.

Palmer, C.T. 1992. "The Northwest Newfoundland fishery crisis: formal and informal management options in the wake of the northern cod moratorium." Institute of Social and Economic Research, report no. 6. St. John's: ISER.

—. 1994. "Folk management, 'Soft evolutionism,' and fishers' motives: implications for the regulation of the lobster fisheries of Maine and Newfoundland." *Human Organization* 52(4):414–20.

—. 1995. "The boats from home: fish smacking tactics and the management of the Labrador fishery." Institute of Social and Economic Research, report no. 22. St. John's: ISER.

Palmer, C.T., and Peter R. Sinclair. Forthcoming. "Perceptions of a fishery in crisis: the attitudes of dragger skippers towards the Gulf of St. Lawrence cod moratorium." *Society and Natural Resources.*

Pearse, Peter H. 1980. "Property Rights and the Regulation of Commercial Fisheries." *Journal of Business Administration* 11(2):185–209.

Pinkerton, Evelyn, ed. 1989. *Co-operative Management of Local Fisheries: New Directions for Improved Management and Community Development.* Vancouver: University of British Columbia Press.

Proskie, John. 1951. *An Appraisal of the French Shore Fishing Industry.* Ottawa: Department of Fisheries and Oceans.

Reimer, Bill. 1992. "Modernization: technology and rural industries and population." In David A. Hoy and Gurcharn S. Basran (eds.), *Rural Sociology in Canada.* Toronto: Oxford University Press.

Reinhardt, Nola, and Peggy Barlett. 1989. "The Persistence of family farms in United

States agriculture." *Sociologia Ruralis* 29:203–25.
Richards, Canon J.T. 1953. "The First settlers on the French Shore." *Newfoundland Quarterly* 52(2):17–19, 44; and 52(4):15–16.
Scott, Anthony. 1955. "The fishery: the objectives of sole ownership." *Journal of Political Economy* 63:116-24.
—. 1979. "Development of economic theory on fisheries regulation." *Journal of the Fisheries Research Board of Canada* 36:725-41.
Scott., W.B., and M.G. Scott. 1988. "Atlantic Fishes of Canada." *Canadian Bulletin of Fisheries and Aquatic Sciences* no. 219.
Sider, Gerald M. 1986. *Culture and Class in Anthropology and History: A Newfoundland Illustration.* New York: Cambridge University Press.
Sinclair, Peter R. 1983. "Fishermen Divided: the impact of limited entry licensing in northwest Newfoundland." *Human Organization* 42(4):307–13.
—. 1985. *From Traps to Draggers: Domestic Commodity Production in Northwest Newfoundland, 1850–1982.* St. John's: Institute of Social and Economic Research.
—. 1986. "The survival of small capital: state policy and the dragger fleet in northwest Newfoundland." *Marine Policy* 10(April):111–18.
—. 1990. "Fisheries management and problems of social justice: reflections on northwest Newfoundland." *Maritime Anthropological Studies* 3(1):30–47.
Sinclair, Peter R., and Lawrence Felt. 1991. "Home, sweet home: dimensions and determinants of life satisfaction in a marginal region." *Canadian Journal of Sociology* 16(1):1–21.
Sinclair, Peter R., with Heather Squires and Lynn Downton. Forthcoming. "A future without fish? Constructing social life on Newfoundland's Bonavista Peninsula after the cod moratorium." In D. Newell and R. Ommer (eds.), *Fishing People, Fishing Places.* Toronto: University of Toronto Press.
Singleton, Sara, and Michael Taylor. 1992. "Common property, collective action and community." *Journal of Theoretical Politics* 4(3):309–24.
Steele, D.H., R. Andersen, and J.M. Green. 1992. "The managed commercial annihilation of northern cod." *Newfoundland Studies* 8:34–68.
Taylor, Lawrence. 1987. "The river would run red with blood: community and common property in an Irish fishing settlement." In B.J. McCay and J.M. Acheson (eds.), *The Question of the Commons.* Tucson: University of Arizona Press.
Thompson, Frederick F. 1961. *The French Shore Problem in Newfoundland.* Toronto: University of Toronto Press.
Thornton, Patricia A. 1977. "The demographic and mercantile bases of initial permanent settlement in the Strait of Belle Isle." In J.J. Mannion (ed.), *The Peopling of Newfoundland.* St. John's: Institute of Social and Economic Research.
Winson, Anthony. 1992. *The Intimate Commodity: Food and the Development of the Agro-industrial Complex in Canada.* Toronto: Garamond.

New Basics from Fernwood Publishing

The Basics present topics of current interest—many on the cutting edge of scholarship—in a short and inexpensive format that makes them ideal as supplementary texts and for general interest. New proposals for the series are welcome.

LAWS, CRIMES AND COMMUNITIES

John L. McMullan, David C. Perrier, Peter D. Swan, Stephen Smith

In this book, McMullan and his colleagues have provided much needed information and analysis on "unconventional" crimes by researching fire for profit, illegal fishing and business crime in Atlantic Canada. The three essays fill an information gap left by scant media reports, conflicting government statistics and, in the case of crimes of capital, wilfully concealed information.

Contents: Going to Blazes: The Social Economy of Arson • Poaching vs. the Law: The Social Organization of Illegal Fishing • Toxic Steel: Corporate Crime and the Contamination of the Environment

104pp Paper ISBN 1 895686 79 2 $12.95

IMMIGRATION AND THE LEGALIZATION OF RACISM

Lisa Marie Jakubowski

"The chameleon-like nature of the law—the duplicitous ways in which the law is written, the equivocal way in which it is stated and, therefore, talked about, the hiding of the truth about the resources which are expended in its implementation, the misleading way in which it casts the discretions it purports to take away and to give—its ideological functioning and its capacity to legitimate the illegitimate, all are put under the microscope in this study. It is a timely piece of work. It may make some readers uncomfortable, but it will leave no one untouched." *—H.J. Glasbeek, Professor Emeritus, Osgoode Hall Law School*

Contents: Controlling Immigration: "Race" and Canadian Immigration Law and Policy Formation • The Question of Social Order: Exploring the Duality of Law • Ideology as Methodology: Documentary Analysis of Canadian Immigration Law • Amending the Canadian Immigration Act: The Live-In Caregiver Program • Amending the Canadian Immigration Act: Bill C-86 • Concluding Remarks

104pp Paper ISBN 1 895686 74 1 $12.95